Keys to adults of the water beetles of Britain and Ireland (Part 1)

(Coleoptera: Hydradephaga: Gyrinidae, Haliplidae, Paelobiidae, Noteridae and Dytiscidae)

Garth N. Foster

The Aquatic Coleoptera Conservation Trust

Laurie E. Friday

Graduate School of Life Sciences,
University of Cambridge

Colour plates prepared by

Jirí Hájek

National Museum – Prague, Czech Republic

Published for the Royal Entomological Society
The Mansion House
Bonehill
Chiswell Green Lane
Chiswell Green
St Albans
AL2 3NS
www.royensoc.co.uk

By the   Field Studies Council
The Annexe
Preston Montford Lane
Shrewsbury
SY4 1DU
www.field-studies-council.org

ISBN: 978 0 901546 93 7

Contents

Dedication

This work is dedicated to Derek Lott.

Abstract

Illustrated keys are provided for the families, genera, species and other taxa of British and Irish Hydradephaga. Colour plates are provided of all the species. The Hydradephaga comprise the Gyrinidae, Haliplidae, Noteridae, Paelobiidae (also known as Hygrobiidae) and the Dytiscidae. Notes include characters for distinguishing the sexes, some basic information on biology and collecting methods, and reviews of distributions in Ireland and Britain, including the Channel Isles. Appended is a key to the families of aquatic Coleoptera.

Acknowledgements

The following are thanked for access to figures that formed the basis of many drawings: Bas Drost, Møgens Holmen, Anders Nilsson, Edward Tranda, and Bernhard van Vondel. Robert Angus is thanked for advice on some species, in particular the *Nebrioporus depressus-elegans* complex, with some information for that complex being drawn from David Shirt's (1983) unpublished Ph.D. thesis. Johannes Bergsten and Anders Nilsson are thanked for access to their unpublished findings concerning *Suphrodytes*. It is probably unwise to attempt to list all of those who have helped in the production of this work, especially as this might include contributors to the recording scheme data-bases used to generate the comments on distribution. However, in the immediate past we would wish to acknowledge those who have critically appraised near-final drafts: David Bilton, Derek Lott, Martin Luff and Brian Nelson. Malcolm Greenwood recently provided feedback based on the responses of his students, and Martin Drake, Magnus Sinclair, Jim Thomas and Mike Tynen have identified problems that needed attention.

Vit Kabourek, of Kabourek Publishing Ltd, is acknowledged for his permission to access many of the plates.

Introduction

A key to the adults of British water beetles (Friday, 1988) was published by the Field Studies Council to fill the gap left by publications of the Ray Society (Balfour-Browne, 1940, 1950, 1958) and the Royal Entomological Society (Balfour-Browne, 1953). As such Friday (1988) covered more groups than Balfour-Browne's treatment but some had to be omitted. Revision of the 1988 work has become essential because of many changes in the fauna and in the classification and taxonomy of water beetles. The list of taxa requiring treatment nears four hundred, potentially unwieldy as a single book. Beetles fall into three suborders Adephaga, Myxophaga and Polyphaga. This part concerns the aquatic members of the Adephaga, the so-called "Hydradephaga", divided into the whirligig beetles (Gyrinidae), the crawling water beetles (Haliplidae), the squeak beetle in the Paelobiidae (also referred to as the Hygrobiidae) and the diving beetles in the Noteridae and Dytiscidae. *The Carabidae (ground beetles) of Britain and Ireland* (Luff, 2007) covers the terrestrial Adephaga (or Geadephaga). Another kind of sister volume will be Part 2 of this work, concerned with the aquatic Polyphaga. The dilemma as to what should be described as an aquatic beetle can be avoided until the second volume as the Hydradephaga, treated here, are aquatic in all stages of their life-cycle except the pupa.

It is hoped that this work will be found useful by limnologists. A key to the families of all wetland beetles is included here (Key 23, Appendix, p. 113) in an attempt to reduce the problems that someone, not necessarily an entomologist, might experience when dealing with an array of Coleoptera for the first time. Otherwise the format of this work follows the rest of the Handbook series. Streamlining for movement in water often effaces potential structural differences whereas colour and body size can be more easily described and seen. Thus the keys do not necessarily follow systematic differences, with some subgroups within keys isolated by use of seemingly trivial characters rather than by orthodox morphology. The descriptions that are here applied to genera may in fact be associated only with the one or two species by which the genus is represented in Britain and Ireland.

Morphology

The most obvious distinguishing feature of adults of the Hydradephaga is that they cannot only swim but also dive. They come to the surface rear end uppermost in order to renew the air supply between the elytra and the abdomen. Most other water beetles cannot swim – they at most walk, upside down, just under the water's surface. Exceptions that can dive are *Berosus*, *Hydrochara*, *Hydrophilus*, a few *Helophorus* and *Laccobius*. The Polyphagan aquatic beetles have extensive parts of their underside covered by a bubble and generally break the surface head first to renew their air supply.

Swimming dictates the body shape of most Hydradephaga, typically streamlined and tapering to the rear. Swimming has had a profound effect on the thorax in that it is the last pair of legs that provides the main thrust when diving. The metacoxae, the coxae of the hind legs, differ from those of the two front pairs of legs, being large and fixed, usually incorporated into the main body shell (Fig. 1). The aquatic beetles in the Polyphaga that swim also have a smooth outline, but the metacoxae are never as strongly differentiated as in the Adephaga.

It should be noted that the expressions "fore", "mid" and "hind" are used interchangeably with both first, second and third and "pro-", "meso-" and "meta-", the latter suffixes often being more convenient to use than the ordinary words.

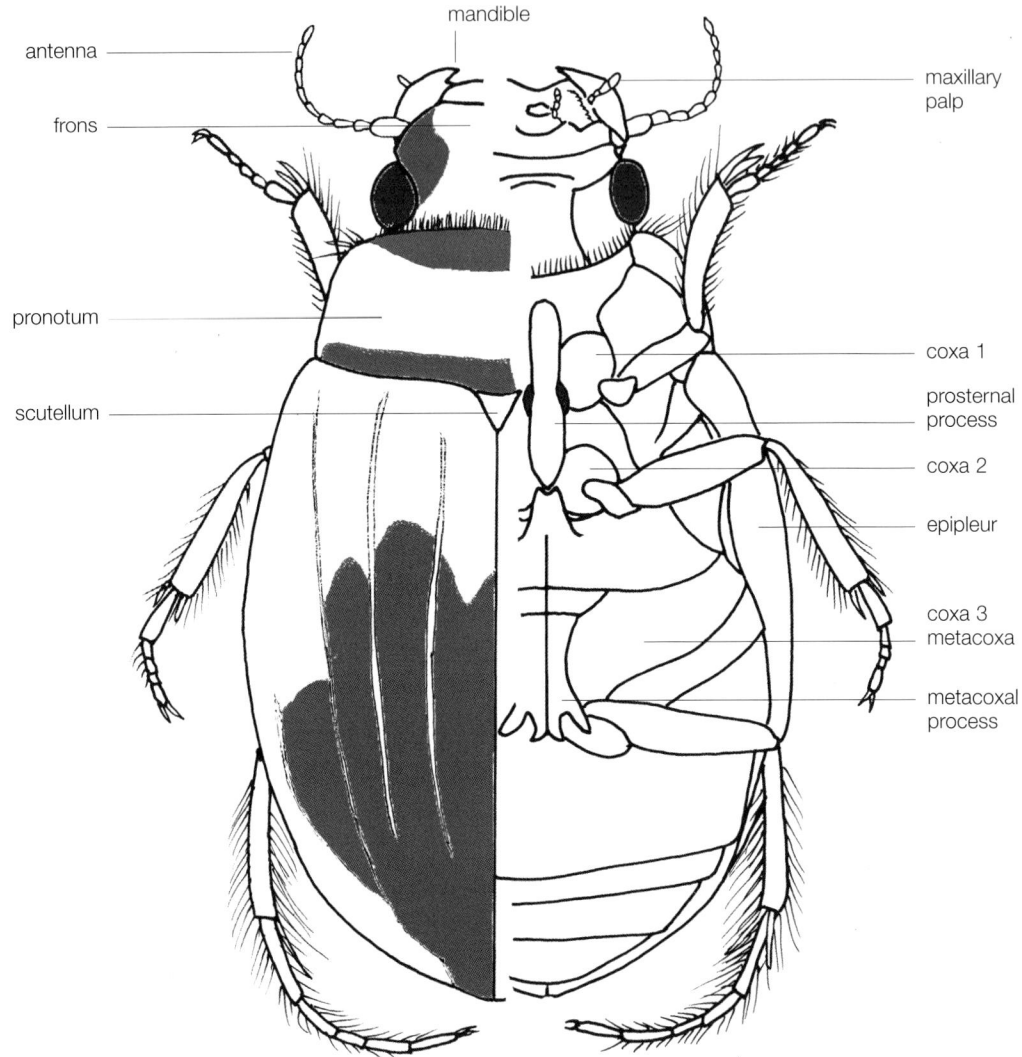

Figure 1. Structure of the squeak beetle, *Hygrobia hermanni*

Other features of the Adephaga are that they are largely predators both as adults and larvae. Most predatory beetles have their mouthparts, with the mandibles most prominent, pointing forwards (the prognathous condition): this is true for many Hydradephaga but the forward sweep of the dome of the head can make this feature not as obvious as in ground beetles. The Polyphaga include predatory species, for example the larvae of the Hydrophiloidea, but the majority are plant, biofilm, and detritus-feeders, with some groups showing a high degree of host plant specialisation. All Adephaga have defence glands that produce toxins to deter fish and other predators: the discharged fluids can produce a clinging pungency varying in character enough to identify genera and even a few individual species.

Detailed morphology will be considered for each family in turn.

Size

Beetles have complete metamorphosis with the egg developing into a larva that usually casts its cuticle three times (hence three instars) before pupating. Once emerged from the pupa the adult is fixed in size. Size thus provides a crucial identification character. If a beetle falls outside the range given for each species that has been identified then it is likely that the identification is wrong. Measurements are from the tip of the elytra to the front of the head.

Collecting methods

The traditional equipment of choice in Britain and Ireland to catch water beetles is the pond net, preferably with a D-framed or circular rim strong enough to push aside vegetation, and with the fabric edge of the net bag protected inside the main rim: the mesh of the bag is typically 1 mm. However, kitchen sieves are popular in other countries and can be just as effective for the smaller beetles. Almost as important as the net or sieve is the sorting tray, which should be large as fieldwork constraints will allow. The first sweep through the water is often the only one likely to capture the larger diving beetles, which are more amenable to capture by trapping. A trap is easily made by cutting off the neck of a 1 litre plastic bottle and securing it back-to-front to form a "minnow trap". Proprietary cat food or similar can be used as a bait. The most important consideration with netting is to throw the sorted debris back into the water. It is crucial that traps are removed after use, and that this use should not be for too long, perhaps no more than overnight if there is to be any chance of reviving unwanted beetles and returning them safely to the water.

Many diving beetles, even the larger ones, will live in extremely shallow water, so working of any vegetated edge is essential. However, haliplids and a few diving beetle species are only to be found in open water.

Checklist

Changes in the nomenclature of the Hydradephaga since Balfour-Browne's works (1940, 1950, 1953) were discussed by Foster (2004). Names here follow slightly more recent treatments, Löbl & Šmetana (2003) for Dytiscidae and Gyrinidae, Nilsson & van Vondel (2005) for the rest. Tribes contribute little to the understanding of the rather limited fauna of these islands, and are not included. The name *Liopterus* has replaced *Copelatus* for the British and Irish species (Balke, Ribera & Vogler, 2004), and the genus *Boreonectes* has been created for one complex of *Stictotarsus* (Angus, 2010). Several dytiscid species have two forms of female, one male-like in its surface features and the other matt or with grooved ("sulcate") elytra. Some of these, e.g. *Hydroporus memnonius* and its variety *castaneus* (see Bilton, Thompson & Foster, 2008), have different distributions and are worth recording as separate entities.

Suborder Adephaga Clairville, 1806

Family GYRINIDAE Latreille, 1810

    Subfamily GYRININAE Latreille, 1810
    GYRINUS O.F. Müller, 1764
        Subgenus GYRINULUS O.F. Müller, 1764
          minutus Fabricius, 1798
        Subgenus GYRINUS O.F. Müller, 1764
          aeratus Stephens, 1835
          caspius Ménétriés, 1832
          distinctus Aubé, 1837
            colymbus auctt.
          marinus Gyllenhal, 1808
          natator (Linnaeus, 1758)
          opacus C.R. Sahlberg, 1819
          paykulli Ochs, 1927
            bicolor auctt.
          substriatus Stephens, 1828
          suffriani Scriba, 1855
          urinator Illiger, 1807
    ORECTOCHILUS Dejean, 1833
        Subgenus ORECTOCHILUS Dejean, 1833
          villosus (O.F. Müller, 1776)

Family HALIPLIDAE Kirby, 1837

    BRYCHIUS C.G. Thomson, 1859
        elevatus (Panzer, 1793)
    HALIPLUS Latreille, 1802
        Subgenus HALIPLIDIUS Guignot, 1928
          confinis Stephens, 1828
          obliquus (Fabricius, 1787)
          varius Nicolai, 1822

Subgenus HALIPLUS Latreille, 1802
   apicalis C.G. Thomson, 1868
   fluviatilis Aubé, 1836
   furcatus Seidlitz, 1887
   heydeni Wehncke, 1875
   immaculatus Gerhardt, 1877
   lineolatus Mannerheim, 1844
   ruficollis (De Geer, 1774)
   sibiricus Motschulsky, 1860
      wehnckei Gerhardt, 1877
Subgenus LIAPHLUS Guignot, 1928
   flavicollis Sturm, 1834
   fulvus (Fabricius, 1801)
   laminatus (Schaller, 1783)
   mucronatus Stephens, 1828
   variegatus Sturm, 1834
Subgenus NEOHALIPLUS Netolitzsky, 1911
   lineatocollis (Marsham, 1802)
PELTODYTES Régimbart, 1879
   caesus (Duftschmid, 1805)

Family NOTERIDAE C.G. Thomson, 1860

Subfamily NOTERINAE C.G. Thomson, 1860
NOTERUS Clairville, 1806
   clavicornis (De Geer, 1774)
      sparsus (Marsham, 1802)
   crassicornis (O.F. Müller, 1776)
      capricornis (Herbst, 1784)
      minor Balfour-Browne, 1962

Family PAELOBIIDAE Erichson, 1837
HYGROBIIDAE Régimbart, 1879

HYGROBIA Latreille, 1804
   hermanni (Fabricius, 1775)
   tarda (Herbst, 1779)

Family DYTISCIDAE Leach, 1813

Subfamily AGABINAE C.G. Thomson, 1867
AGABUS Leach, 1817
Subgenus ACATODES C.G. Thomson, 1859
   arcticus (Paykull, 1798)
   congener (Thunberg, 1794)
   sturmii (Gyllenhal, 1808)
Subgenus AGABUS Leach, 1817
   labiatus (Brahm, 1791)
   uliginosus (Linnaeus, 1761)
      male-like form
      matt var. *dispar* (Bold, 1849)
   undulatus (Schrank, 1776)

Subgenus GAURODYTES C.G. Thomson, 1859
 affinis (Paykull, 1798)
 biguttatus (Olivier, 1795)
 bipustulatus (Linnaeus, 1767)
 brunneus (Fabricius, 1798)
 conspersus (Marsham, 1802)
 didymus (Olivier, 1795)
 guttatus (Paykull, 1798)
 melanarius Aubé, 1837
 nebulosus (Forster, 1771)
 paludosus (Fabricius, 1801)
 striolatus (Gyllenhal, 1808)
 unguicularis (Thomson, 1867)
ILYBIUS Erichson, 1832
 aenescens C.G. Thomson, 1879
 ater (De Geer, 1774)
 chalconatus (Panzer, 1796)
 fenestratus (Fabricius, 1781)
 fuliginosus (Fabricius, 1792)
 guttiger (Gyllenhal, 1808)
 montanus (Stephens, 1828)
  melanocornis (Zimmermann, 1915)
 quadriguttatus (Lacordaire, 1835)
 subaeneus Erichson, 1837
 wasastjernae (C.R. Sahlberg, 1824)
PLATAMBUS C.G. Thomson, 1859
 maculatus (Linnaeus, 1758)
  var. *inornatus* Shilsky, 1888
Subfamily COLYMBETINAE Erichson, 1837
COLYMBETES Clairville, 1806
 fuscus (Linnaeus, 1758)
RHANTUS Dejean, 1833
Subgenus NARTUS Zaitzev, 1907
 grapii (Gyllenhal, 1808)
Subgenus RHANTUS Dejean, 1833
 bistriatus (Bergsträsser, 1778)
  aberratus Gemminger & von Harold, 1868
  adspersus sensu Balfour-Browne, 1950
 exsoletus (Forster, 1771)
 frontalis (Marsham, 1802)
 suturalis (Macleay, 1825)
  pulverosus (Stephens, 1828)
 suturellus (Harris, 1828)
  bistriatus sensu Balfour-Browne, 1950
Subfamily COPELATINAE Branden, 1885
LIOPTERUS Dejean, 1833
COPELATUS Erichson, 1832
 haemorrhoidalis (Fabricius, 1787)

Subfamily DYTISCINAE Leach, 1815

    ACILIUS Leach, 1817

        canaliculatus (Nicolai, 1822)

        sulcatus (Linnaeus, 1758)

    GRAPHODERUS Dejean, 1833

        bilineatus (De Geer, 1774)

        cinereus (Linnaeus, 1758)

        zonatus (Hoppe, 1795)

    CYBISTER Curtis, 1827

        lateralimarginalis (De Geer, 1774)

    DYTISCUS Linnaeus, 1758

        circumcinctus Ahrens, 1811

            sulcate form

            male-like form *flavocinctus* Hummel, 1823

        circumflexus Fabricius, 1801

            male-like form

            sulcate form *dubius* Audinet-Serville, 1830

        dimidiatus Bergsträsser, 1778

        lapponicus Gyllenhal, 1808

        marginalis Linnaeus, 1758

        semisulcatus O.F. Müller, 1776

    HYDATICUS Leach, 1817

        seminiger (De Geer, 1774)

        transversalis (Pontoppidan, 1763)

Subfamily HYDROPORINAE Aubé, 1836

    BIDESSUS Sharp, 1882

        minutissimus (Germar, 1824)

        unistriatus (Goeze, 1777)

    HYDROGLYPHUS Motschulsky, 1853

        GUIGNOTUS Houlbert, 1934

        geminus (Fabricius, 1792)

            pusillus (Fabricius, 1781)

    DERONECTES Sharp, 1882

        latus (Stephens, 1829)

    GRAPTODYTES Seidlitz, 1887

        bilineatus (Sturm, 1835)

        flavipes (Olivier, 1795)

        granularis (Linnaeus, 1767)

        pictus (Fabricius, 1787)

    HYDROPORUS Clairville, 1806

        angustatus Sturm, 1835

        discretus Fairmaire & Brisout de Barneville, 1859

        elongatulus Sturm, 1835

        erythrocephalus (Linnaeus, 1758)

            nominate form

            form *deplanatus* Gyllenhal, 1826

        ferrugineus Stephens, 1829

        glabriusculus Aubé, 1838

        gyllenhalii Schiødte, 1841

            gyllenhali unjustified emendation

incognitus Sharp, 1869
longicornis Sharp, 1871
longulus Mulsant & Rey, 1861
marginatus (Duftschmid, 1805)
melanarius Sturm, 1835
memnonius Nicolai, 1822
    male-like form
    form *castaneus* Aubé, 1838
morio Aubé, 1838
necopinatus Fery, 1999
    cantabricus by usage, not Sharp, 1882
neglectus Schaum, 1845
nigrita (Fabricius, 1792)
obscurus Sturm, 1835
obsoletus Aubé, 1838
palustris (Linnaeus, 1761)
planus (Fabricius, 1782)
pubescens (Gyllenhal, 1808)
    melanocephalus (Marsham, 1802)
rufifrons (O.F. Müller, 1776)
scalesianus Stephens, 1828
striola (Gyllenhal, 1826)
tessellatus (Drapiez, 1819)
    tesselatus misspelling
tristis (Paykull, 1798)
umbrosus (Gyllenhal, 1808)
NEBRIOPORUS Régimbart, 1906
    assimilis (Paykull, 1798)
    canaliculatus (Lacordaire, 1835)
    depressus (Fabricius, 1775)
    elegans (Panzer, 1794)
OREODYTES Seidlitz, 1887
    alpinus (Paykull, 1798)
    davisii (Curtis, 1831)
        davisi unjustified emendation
    sanmarkii (C.R. Sahlberg, 1826)
        sanmarki unjustified emendation
    septentrionalis (Gyllenhal, 1826)
PORHYDRUS Guignot, 1945
    lineatus (Fabricius, 1775)
SCARODYTES des Gozis, 1914
    halensis (Fabricius, 1787)
STICTONECTES Brinck, 1943
    lepidus (Olivier, 1795)
STICTOTARSUS Zimmermann, 1919 sensu Angus, 2010
    duodecimpustulatus (Fabricius, 1792)
BOREONECTES Angus, 2010
    multilineatus (Falkenström, 1922)
        griseostriatus auctt.

SUPHRODYTES des Gozis, 1914
> *dorsalis* (Fabricius, 1787)
> *figuratus* (Gyllenhal, 1826)

HYDROVATUS Motschulsky, 1853
> *clypealis* Sharp, 1876
> *cuspidatus* (Kunze, 1818)

HYGROTUS Stephens, 1828
> Subgenus HYGROTUS Stephens, 1828
> > *decoratus* (Gyllenhal, 1810)
> > *inaequalis* (Fabricius, 1777)
> > *quinquelineatus* (Zetterstedt, 1828)
> > *versicolor* (Schaller, 1783)
> Subgenus COELAMBUS C.G. Thomson, 1860
> > *confluens* (Fabricius, 1787)
> > *impressopunctatus* (Schaller, 1783)
> > > male-like form
> > > matt var. *lineellus* Gyllenhal, 1808
> > *nigrolineatus* (von Steven, 1808)
> > > *lautus* (Schaum, 1843)
> > *novemlineatus* (Stephens, 1829)
> > *parallellogrammus* (Ahrens, 1812)
> > > *parallelogrammus* unjustified emendation

HYPHYDRUS Illiger, 1802
> *aubei* Ganglbauer, 1892 [Channel Islands only]
> *ovatus* (Linnaeus, 1761)

LACCORNIS des Gozis, 1914
> *oblongus* (Stephens, 1835)

Subfamily LACCOPHILINAE Gistel, 1856
LACCOPHILUS Leach, 1815
> *hyalinus* (De Geer, 1774)
> *minutus* (Linnaeus, 1758)
> *poecilus* Klug, 1834
> > *obsoletus* by usage, not Westhoff, 1881
> > *ponticus* Sharp, 1882
> > *variegatus* (Germar & Kaulfuss, 1816)

Common names

Some of the choicer common names are included in the individual species accounts for those who might wish to use them. It should, however, be understood that their use in the absence of a Latin name will cause confusion. They are not in common parlance, as for example is the case for Odonata and the larger Lepidoptera. The Law of Priority does not apply to common names, a preferred name coming about by usage. A few of those listed here are recently conceived but most are drawn from older publications.

Keys to families

A key to families of aquatic beetles is provided in Appendix 1 (page 113), along with guidance on water beetles for the non-specialist.

Keys are organised by family with keys to genus and species.

Species keys are followed by species notes which should be referred to for identification confirmation.

Distributions are given on pages 109 and 110. Colour plates start on page 123.

Family GYRINIDAE Whirligig beetles

This is one of the most distinctive families of beetles as a whole, well known for their surface-swimming behaviour and for having two pairs of eyes, one for seeing in the water and one for seeing in air. Despite their sleek appearance whirligigs are one of the most ancient groups of beetle, being recognisable as fossils back to the Jurassic. In other Hydradephaga the hind coxae (metacoxae) are modified into fixed plates but in the Gyrinidae this is also the case for the middle coxae (mesocoxae) (Fig. 3). In the other Hydradephaga the main thrust is given by the hind legs whereas whirligigs use mainly the second pair of legs, the third being deployed to accelerate. Whirligigs are generally regarded as predators of insects falling onto the surface of the water, and may occur in large groups (flotillas or schools) during the day, though some are active at night. The larvae are also predatory: they breathe by means of tracheal gills developed as abdominal processes, and this may well explain the association of most species with well-oxygenated water in lakes and running water.

The tips of the genitalia usually protrude from the last abdominal segment sufficiently to differentiate the male three-pronged structure, with an aedeagus flanked by parameres, from the two vulval sclerites of the female. Gentle squeezing of a freshly dead specimen will usually extrude the rest, but the contents of the abdomen quickly autolyse, when squeezing results in the rear end breaking off. The aedeagus of the male is generally diagnostic, its length relative to the parameres being useful. Gyrinidae are one of the few groups of beetle in which the female genitalia differ sufficiently from one species to another to be diagnostic. Females are often larger than males, so much so in *G. marinus* that the sexes are often mistaken for different species in the field.

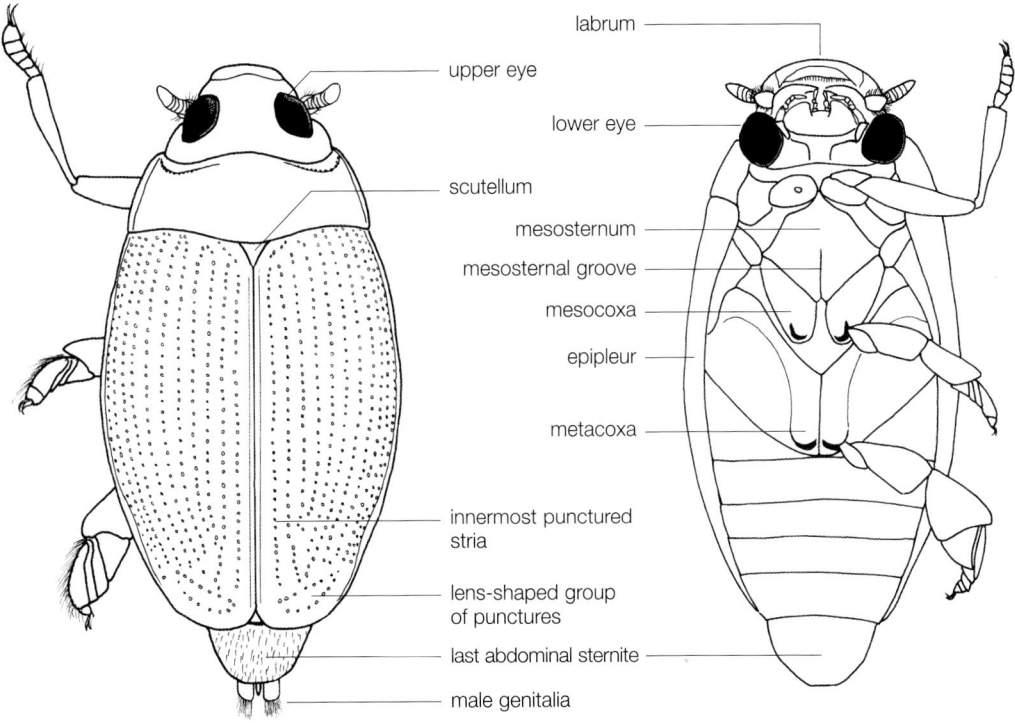

Figure 2. *Gyrinus marinus*
(based on Holmen 1987)

Figure 3. *Gyrinus distinctus*
(based on Holmen 1987)

Key 1. The genera of Gyrinidae

1. Pronotum and elytra without hair; labrum broad and forming a continuous curve with the head; edges of pronotum and much of elytra visible from above (arrowed, Fig. 4) because the body is fairly flat; underside black, reddish or orange .... 1. *Gyrinus* Linnaeus (below)

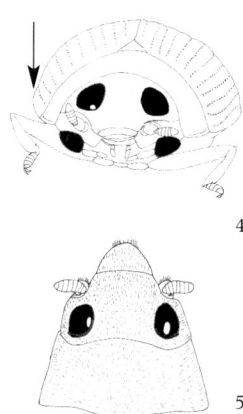

4

- Pronotum and elytra covered in fine hair (Fig. 5); labrum protruding, not forming a smooth outline with the rest of the head; body laterally compressed such that the edges of the pronotum and elytra are not visible from above (arrowed, Fig. 6); underside mainly orange .... 2. *Orectochilus* Dejean (p. 20)

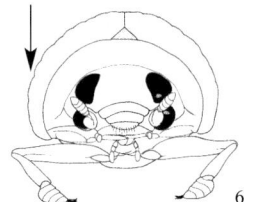

5

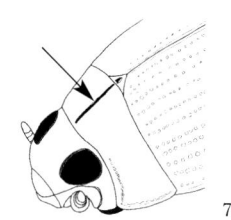

6

1. *GYRINUS* Linnaeus

Key 2. The species of *Gyrinus*

1. Scutellum (the triangular plate at the junction of pronotum with elytra) with a short longitudinal ridge and pronotum with a median longitudinal "seam" running its length (Fig. 7), a linear gap in the microreticulation that covers most of the upper side (view obliquely to be certain of seeing these features); mesosternum (between middle and front coxae) with a central groove running its full length (Fig. 8); underside yellow or orange; genitalia as illustrated (Figs 9 and 10) but dissection should not be necessary .... Subgenus *Gyrinulus* O.F. Müller .... 1. *Gyrinus minutus* Fabricius (p. 17)

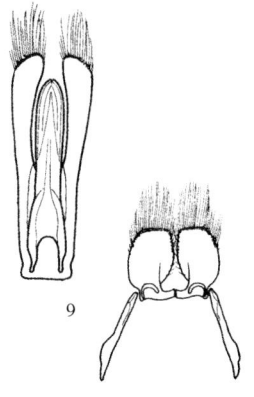

7

9

8

10

- Scutellum flat and pronotum without a median seam; mesosternum without a central groove running full length (though a short groove may be present on the rear part); underside yellow, orange or red and black .... Subgenus *Gyrinus* O.F. Müller 2

2. Underside entirely orange; elytra with stripes of metallic green and blue (rock the beetle to let these stripes catch the light); genitalia as illustrated (Figs 11 and 12) but dissection should not be necessary .... 11. *Gyrinus urinator* Illiger (p. 20)

- Underside mainly dark, with some parts often reddish 3

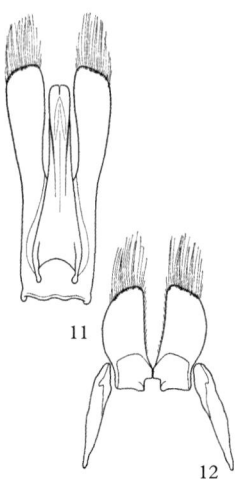

11

12

3. Middle and hind tarsal claws black, darker than the yellow legs and tarsi 4

- Middle and hind tarsal claws yellow, same as the legs 5

4. Aedeagus of male genitalia needle-sharp (Fig. 13); as viewed from above, flattened lateral rim of the elytra broadening out behind the widest point of the elytra; female genitalia (Fig. 14) ....
.... 5. *Gyrinus marinus* Gyllenhal (p. 18)

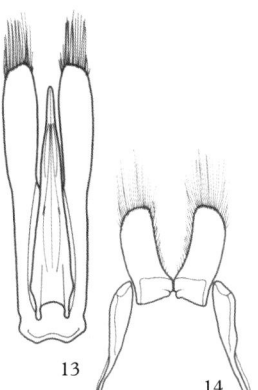

13

14

- Aedeagus blunt at tip (Fig. 15); rim of elytra not widening out, if anything narrowing slightly behind the widest point of the elytra; female genitalia (Fig. 16) .... 2. *Gyrinus aeratus* Stephens (p. 17)

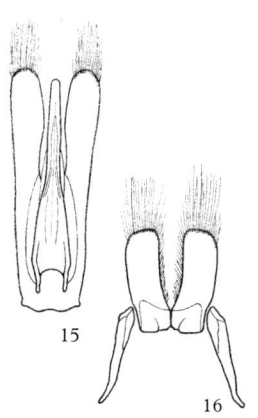

15

16

14

5. Pronotum and elytra dull all over and clearly reticulated when viewed at x 20; underside generally dark all over except for the underside parts of the elytra, the epipleurs, which may also be dark; genitalia (Figs. 17 and 18) blackened at extremities .... 7. *Gyrinus opacus* C.R. Sahlberg (p. 19)

- Pronotum and elytra shining, with at most weak reticulation; underside usually with the mesosternum and/or the last abdominal segment orange in addition to the epipleurs; genitalia pale throughout, rarely darkened 6

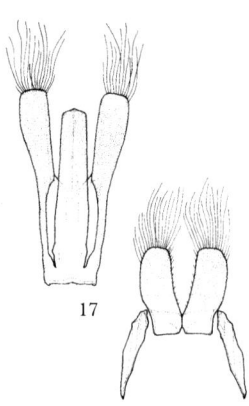

6. Elongate ("canoe-shaped") beetles with elytra scarcely wider at their mid-point than at their front margin (Fig. 19) 7

- Rounded ("dinghy-shaped") beetles; elytra about 1¼ times wider at widest point than at the front margin (Fig. 20) 8

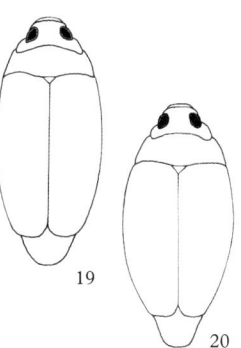

7. Tip of aedeagus broad, like a spatula (Fig. 21), about ⅔ rds the width of the parameres, the outer lobes; female genitalia with ends of the lobes sloping outwards (Fig. 22); mesosternum orange or orange-red .... 3. *Gyrinus caspius* Ménétriés (p. 17)

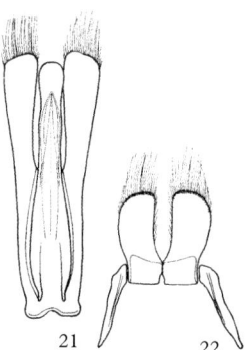

- Tip of aedeagus narrow (Fig. 23), less than a ¼ the width of the parameres; female genitalia with ends of the lobes sloping inwards (Fig. 24); mesosternum black .... 8. *Gyrinus paykulli* Ochs (p. 19)

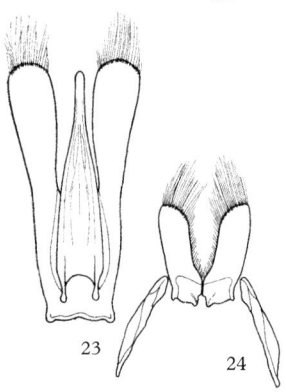

8. Lens-shaped group of punctures at the tip of each elytron (see Fig. 2) fainter than other punctures; aedeagus narrower than and almost as long as the parameres (Fig. 25); female genitalia (Fig. 26) .... 10. *Gyrinus suffriani* Scriba (p. 19)

- Lens-shaped group of punctures at the tip of each elytron deeply impressed, larger and stronger than other punctures; aedeagus narrow or broad (Figs 27, 30 and 34), shorter than the parameres 9

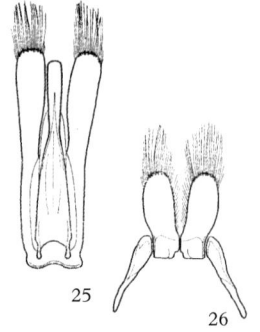

25

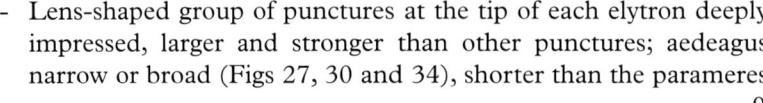

26

9. Aedeagus with an expanded and bluntly pointed tip almost as wide as one of the parameres (Fig. 27); female genitalia about twice as long as broad (Fig. 28); elytra with many dense and fine micropunctures (seen at × 40), about 8 between each of the coarsely punctured rows .... 4. *Gyrinus distinctus* Aubé (p. 18)

- Aedeagus narrowed to about ½ the width of a paramere and truncated, i.e. "cut off", at the tip; female genitalia 3 or more times as long as broad (Figs 31 and 35); elytra with very fine micropunctures (not visible at × 40), 2-3 between each punctured row 10

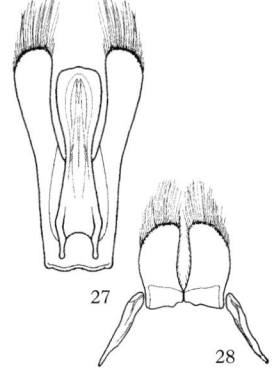

27

28

10. Side of pronotum with a short lateral groove (Fig. 29); female genitalia with a process at the point on the base of each vulval sclerite (Figs 31 and 32), requiring extraction of the structure and removal of any potentially obscuring tissue; mesosternum black; last abdominal sternite usually black; elytra with all punctured rows of about the same intensity .... 6. *Gyrinus natator* (Linnaeus) (p. 18)

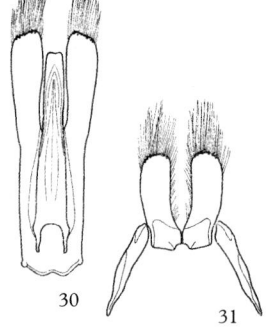

30

31

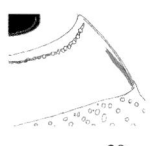

29

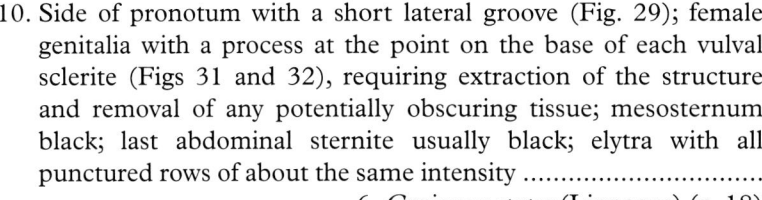

32

- Side of pronotum with a long lateral groove (Fig. 33); female genitalia without a distinctive structure at the base (Fig. 35); mesosternum and last abdominal sternite orange or brown; elytra with innermost elytral stria weakly punctured and rest also often weak .... 9. *Gyrinus substriatus* Stephens (p. 19)

33

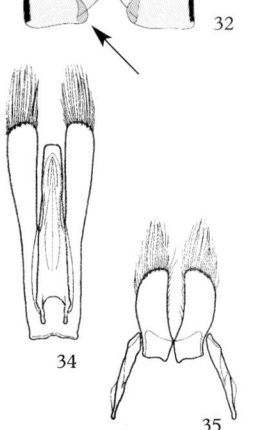

34

35

Subgenus *Gyrinulus* O.F. Müller

1. *Gyrinus minutus* Fabricius Plate 1

Length 3.5-4.5 mm. With practice the smallest whirligig can be detected in the field as its ripples are weaker than those created by other species. Once caught, its size and orange underside are diagnostic in the field. This species will dive when disturbed but more often individuals break away from flotillas of other species with which it mixes and make the nearest equivalent of a whirligig straight line escape. It is confined to still water, usually on peat in dubh lochan complexes, and also on western lakes. This species has died out over much of England, still being found on remnant heathland in Dorset and the Lake District, with relatively modern records for South Essex in Epping Forest, and south-west Yorkshire at the Askern Pool. In Wales this species is now confined to Snowdonia. Known from Arran, Barra, Berneray, Coll, Cumbrae, Eigg, Fladda, Islay, Jura, Lewis and Harris, Mingulay, Muck, Mull, Raasay, Rum, Scalpay, Shetland, Skye, South Rona, South and North Uist. Frequent in the north and west of Ireland, reaching Achill, Clare, Inishbofin, Rathlin, and Tory Islands. There are old records for the Isle of Man and Jersey. Reported February-December, peaking in July and September.

Subgenus *Gyrinus* O.F. Müller

2. *Gyrinus aeratus* Stephens Plate 2

Length 4.5-6.3 mm. A small and dark species of open water on lakes, canals and larger drains and slow parts of rivers, usually seen in large flotillas, sometimes mixed with species such as *G. marinus*. These whirligigs swim out of reach rather than diving. The blunt tip of the aedeagus will distinguish this species from *G. marinus*, the point often being visible without the need for dissection. Scattered across England with modern records for the Somerset Levels, Dorset, West Sussex, East Kent, Oxfordshire, East and West Norfolk, Leicestershire, Hatchmere in Cheshire, and the Yorkshire Wolds. The only modern Welsh record is from the Montgomery Canal. In Scotland the species is frequent in Dumfries and Galloway, some Renfrewshire lochs, in Speyside, and lochs in South Aberdeenshire: islands with records are Barra, Colonsay, Eriskay, Islay, Lewis, Raasay, Skye, Soay, South and North Uist. *G. aeratus* is frequent over much of the north and west of Ireland, in the south known from Kerry and Waterford. There are old records for Jersey. Reported January-November, peaking in June and September.

3. *Gyrinus caspius* Ménétriés Plate 3

Length 5.0-7.5 mm. Usually found in groups in amongst emergent vegetation in ponds, ditches and canals, tending to stay afloat and highly active when disturbed. The narrow shape of *G. caspius* might be confused with that of the darker *G. paykulli*. This species has an unusual distribution, being coastal over much of its range in England, Wales and Scotland, often in brackish water, but also occurring in the Border Mires at up to 330 metres above sea level, also on Speyside and a few other inland locations. The northernmost records are for Handa Island, Sutherland and for Lewis. *G. caspius* is also most common in Ireland on the coast but it is found inland, being particularly frequent in the interdrumlin mires of Armagh and Down. Known from Alderney, Guernsey and Jersey. Reported January-November, peaking in May and August.

4. *Gyrinus distinctus* **Aubé** Plate 4

Length 5.0-7.0 mm. In groups but often caught in the net before being seen on the surface, keeping within marginal, emergent vegetation. The shape of this species is intermediate between the "canoe" of *G. caspius* and *G. paykulli* and the "dinghy" of other species. The micropunctured surface confers a bluish bloom on this species, but it is best distinguished by dissection of the males to reveal the broad aedeagus. Until recently it was known as a species of thin vegetation in wind-exposed lochs in Argyll, Kirkcudbrightshire, Islay and Mull, and also slow-flowing waters and large pools, drains and rivers in the Broads, also in East Kent, in Cheshire, and with old records for the Lake District in Blelham Tarn and Derwent Water. However it has now colonised larger post-industrial waters, such as quarry and brickpit ponds, subsidence flashes, and canals in the English Midlands. In Wales it is known only from the Monmouthshire levels. *G. distinctus* is frequent in the western half of Ireland, particularly in the Lough Erne complex in Fermanagh. Reported February-December, peaking in May and July. This species was once mistakenly known as *Gyrinus colymbus* in Britain: *G. colymbus* is known here only from a Mediaeval deposit from Leicestershire.

5. *Gyrinus marinus* **Gyllenhal** Plate 5

Length 4.5-7.5 mm. Although this is the commonest species of open waters on base-rich lakes and large ponds, often forming huge flotillas that move out of reach, it has a restricted distribution. Females are larger than the males giving the impression that two species are involved. The species with which it is most likely to be mixed is *G. aeratus*, the aedeagal tip of which is produced into a blunt point as opposed to the sharp one of *G. marinus*. Despite its name *G. marinus* is not particularly associated with coastal waters. It is most frequent in low-lying fen areas of central and southern England. It appears to have died out in north-east England and is coastal in the north-west. In Scotland it is frequent in the loch-rich Dumfries and Galloway and is known north to the Black Isle, but is absent from much of Highlands and has not been found on any Scottish island except Cumbrae. There are records for Anglesey, the Isle of Man, and the Isle of Wight. Welsh records are mainly coastal except for a strong incursion from off the Cheshire Plain into Flint and Montgomeryshire. *G. marinus* is common over much of north and central Ireland, with very few records from the south. Reported throughout the year, peaking in May and August.

6. *Gyrinus natator* **(Linnaeus)** Plate 6

Length 4.5-6.1 mm. This species was much confused with *G. substriatus* in the past, and many popular accounts wrongly apply the name *natator*. It has been recorded with certainty in Britain only in Newton Reigny Moss and Cliburn Moss in Cumbria up to 1921, both peat cutting sites that have subsequently become overgrown. It occurs in partly shaded pools in association with lakes and large peat cuttings over much of lowland Ireland, usually skulking in ones and twos rather than in large flotillas. The dark underside provides the best determinant in the field, this being one of the few water beetles where the female genitalia provide a more definitive character than the male parts. The inside edges of the vulval sclerites must be cleaned to see the processes, which vary in development but are more distinct than in *G. substriatus*. Reported April-October, peaking in June and August.

7. *Gyrinus opacus* **C.R. Sahlberg** Plate 7

Length 5.0-6.5 mm. There was confusion in the past between this species and *G. aeratus* and *G. marinus*. *G. opacus* appears a little dull in the field and might be mistaken for a large *G. minutus*, a species with which it often occurs. Microscopic examination will reveal the microreticulation, and the species has appendages other than the claws darkened. Confined to Highland Scotland, the Flow Country and Harris, this species is rarely seen on the surface but it can be flushed in numbers from beneath undercut banks in exposed peat lochans. Reported April-October, peaking in June and September.

8. *Gyrinus paykulli* **Ochs** Plate 8

Length 5.5-7.8 mm. This species typically skulks in reedbeds in lakes and can occur in base-enriched sites. Flotillas stage a high speed "dance" when disturbed but usually under cover. The narrow body is more parallel-sided than in *G. caspius*, and it is generally larger and darker than that species with a pointed as opposed to blunt apex to the aedeagus. The English distribution is mainly eastern from Sussex to Yorkshire, with scattered modern records from Bedfordshire, Oxfordshire, Cheshire, and the Cumberland coast. In Wales it is known only from Anglesey, and there are two confirmed Scottish sites in Fife. In Ireland it occurs in a broad band from Down to Clare, mainly on the karst. Known from Jersey. Reported throughout the year except February, peaking in May but common from then until and including October. This species used to be known as *G. bicolor*.

9. *Gyrinus substriatus* **Stephens** Plate 9

Length 5.0-7.0 mm. The common whirligig can be found on almost any man-made pond and ditch, shaded forestry fire ponds being particularly favoured, but it is replaced by *G. marinus* in larger ponds. The name *substriatus* owes to the punctured striae next to the suture being more weakly punctured than the rest: this character is not wholly reliable and some *G. substriatus* can have all of the striae similarly weak, with the surface coarsely scratched: *G. natator* on the other hand always has the striae next to the suture as strong as the rest. However, the longer groove on the side of the pronotum, combined with some parts of the underside being red, are a more reliable guide to distinguish this species from *G. natator*. The distribution covers the whole of Britain and Ireland including almost every island except the Scillies. Known from Guernsey and Jersey. Reported throughout the year, peaking in July, and active even in very cold weather.

10. *Gyrinus suffriani* **Scriba** Plate 10

Length 4.0-6.2 mm. This is the rarest surviving whirligig in England, largely confined to relict pools and grazing fen ditches. It gives the impression of being the fastest *Gyrinus* species, outstripped only by *Orectochilus villosus*, and it is usually found singly in the edges amongst vegetation. Its speed and small size should alert to its presence in the field, the male genitalia being definitive. In England it has only ever been known from seven vice-counties, with comparatively modern records from Berkshire, East and West Norfolk, and East Sussex. It is extinct in Scotland, and found in Wales only on Anglesey and in the Crymlyn Bog. Known from Jersey. Reported March-November, peaking in May and September.

11. *Gyrinus urinator* **Illiger** The Artist Plate 11

Length 5.0-7.8 mm. This species has a distinctive orange underside, shared only with the much smaller *G. minutus* and with *Orectochilus villosus*. The elytral punctured lines are iridescent. This is mainly a running water species, with modern records suggesting an expansion in range in England and Wales. It occurs on the Isle of Man but the old record for Scotland is not valid, the northernmost record still being by Balfour-Browne in 1947 in the River Esk in Cumberland. *G. urinator* is now widely distributed in Ireland, known from Armagh and West Donegal in the north to most southern coastal vice-counties. Known from Jersey. Found throughout the year, peaking in May and September.

2. *ORECTOCHILUS* Dejean

1. *Orectochilus villosus* **(O.F. Müller)** The Hairy or Nocturnal Whirligig Plate 12

Length 5.5-6.5 mm. In permanent rivers and wave-washed lake shores, under rocks or amongst overhanging vegetation reaching the water, or in other places affording shelter during the day. This species is active at night, but may be caught in the net almost by accident when disturbed from daytime resting places. *O. villosus* is found across most of Britain and Ireland The Scottish distribution runs to Caithness with the islands occupied being Arran, Bute, Coll, Islay, Lismore, Mull, Skye, South and North Uist. It is also found on Anglesey and the Isle of Man, but neither on the Isle of Wight nor on the Scillies. Found throughout the year, peaking in April and July.

Family HALIPLIDAE Crawling water beetles

The crawling water beetles have been so named because the adults swim by alternating leg movements in a "crawl", and hence they wobble slightly rather than swimming in a straight line like the diving beetles. All are small insects with unusually large punctures, straw-coloured and often with dark markings. The most distinctive structural features are the large metacoxal plates covering much of the hind legs (Fig. 37). The femora are constrained to work between them and the body, perhaps thus improving the swimming action. It has been claimed that the extra bubbles held by these plates improve their ability to survive under water but it is more likely that they provide a buoyancy aid. The adults and larvae are mainly algivorous, though adults will take other sedentary food items such as Hydrozoa and freshwater sponges. Some species appear to be exclusively associated with stoneworts (Characeae) and others facultatively so.

The male genitalia provide important aids to identification. They lie sideways inside the last few abdominal segments and can be extruded by applying gentle, persistent pressure to the freshly killed insect: otherwise one can push the aedeagophore out by inserting a probe into the cavity of the abdomen once it is removed from the body. The aedeagus should be viewed lying on its side. The parameres differ in appearance. The larger one, with a hairy edge and a tuft of hairs at its tip, is useful in distinguishing some species, but the gaps between the two hair patches is more variable than is sometimes portrayed. The larger paramere is sometimes referred to as the left paramere but here it is referred to as the right following Sharp and Muir (1912) and Kelly and Nilsson (2003). The other paramere and the female genitalia are of little use in identification. Males can be distinguished from females by having tufts of sucker hairs on the widened front and middle tarsi. The male fore claws are often unequal in size.

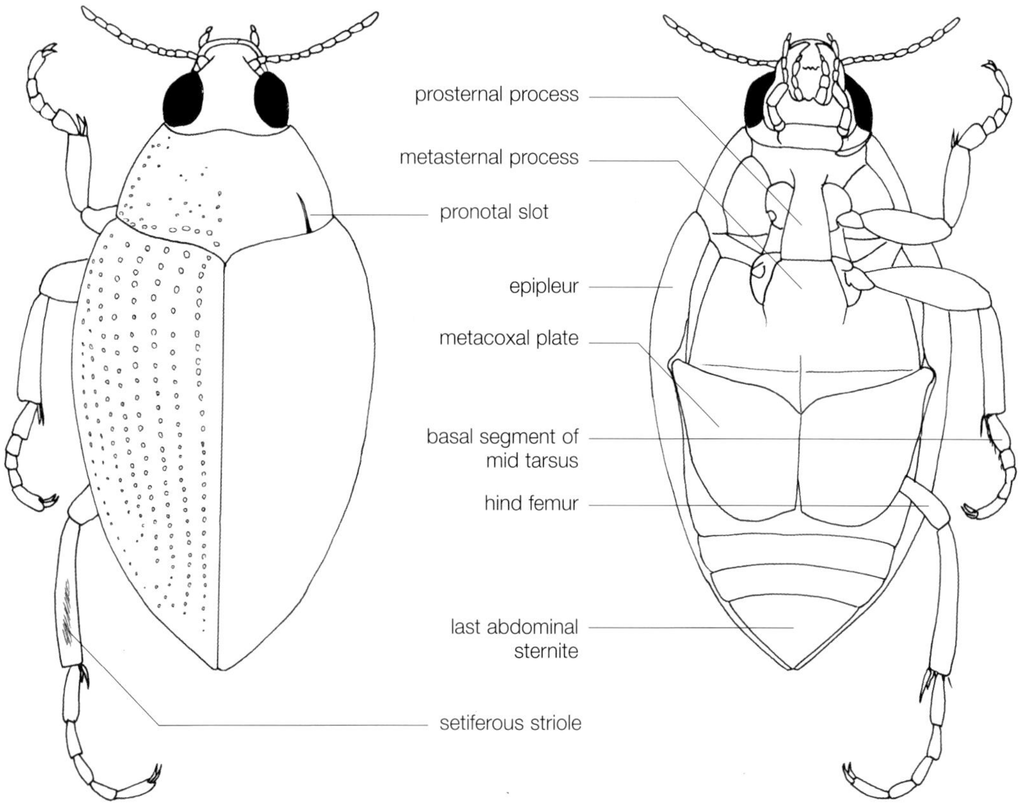

prosternal process

metasternal process

pronotal slot

epipleur

metacoxal plate

basal segment of
mid tarsus

hind femur

last abdominal
sternite

setiferous striole

Figure 36. *Haliplus* upper side features

Figure 37. *Haliplus* underside features

Key 3. The genera of Haliplidae

1. Strong longitudinal ridges on the elytra (Fig. 38); pronotum
   bulging out behind the anterior margin ....
   .... 1. *Brychius* C.G. Thomson (p. 22)

- No protruding ridges on the elytra; pronotum tapering towards
   the anterior margin 2

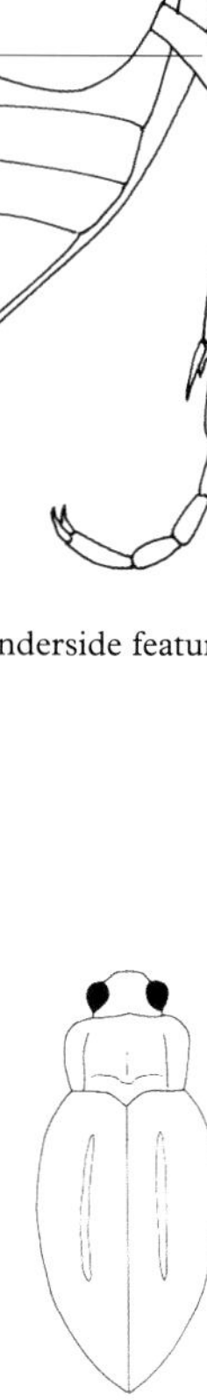

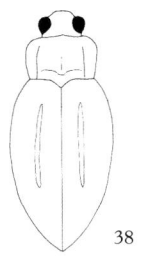

38

2. Hind coxal plates with a smoothly rounded hind margin, leaving
   last three abdominal segments freely visible (Fig. 39); last segment
   of maxillary palp (the longer pair of palps) shorter than the
   preceding one .... 2. *Haliplus* Latreille (p. 22)

- Hind coxal plates each with a point on the hind margin, and
   covering part of the 6th abdominal segment (Fig. 40); last
   segment of maxillary palp longer than the preceding one ....
   .... 3. *Peltodytes* Régimbart (p. 30)

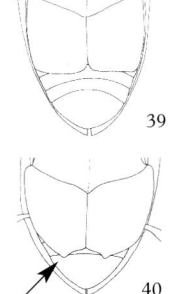

39

40

1. *BRYCHIUS* C.G. Thomson

1. *Brychius elevatus* (Panzer) Plate 13

Length 3.5-4.4 mm. A distinctive species, found in slow running water in canals, rivers, and streams, usually on soft substrata. Widely distributed species through Britain and Ireland, sparsely recorded from the Highlands, and known from Anglesey, the Isle of Man, Islay and the Orkneys. Reported throughout the year, peaking in August.

2. *HALIPLUS* Latreille

Key 4. The species of *Haliplus*

1. Pronotum with a long, curved and pigmented slot on each side; puncture rows at the front edge of the elytra developed into raised ridges; midline usually with a dark line or at least a brown smudge .... Subgenus *Neohaliplus* Netolitzsky .... 17. *Haliplus lineatocollis* (Marsham) (p. 29)

- Pronotum without slots or the slots no more than a third the length of the pronotum; no ridges on the front of the elytra (though the punctures may be enlarged); no central markings on the pronotum 2

2. Dorsal and ventral sides of body with fine punctures clearly visible at × 30 in both sexes, these punctures in addition to the coarser primary punctures; epipleurs without coarse punctures .... Subgenus *Haliplidius* Guignot 3

- Upper surface smooth with at most a partial cover of very fine puncturation visible at × 50, more widely distributed in females of some species; epipleurs with coarse punctures the same size as those of the metacoxal plates (Fig. 41) 5

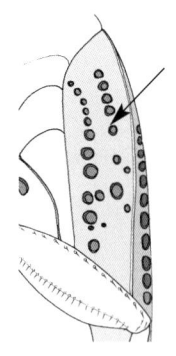

41

3. Smaller species, 2.9 mm or less; pronotum with a broad black band along the hind margin and another along the front margin associated with a black mark on the back of the head; elytra with a speckled pattern; aedeagus with a narrow tip (Fig. 42); paramere (Fig. 43) .... 3. *Haliplus varius* Nicolai (p. 26)

- Larger species, more than 3 mm long; pronotum without black bands, although some diffuse dark pigment and the shadow of the head may be visible along the front edge; aedeagus with a blunt (Fig. 46) or toothed tip (Fig. 44) 4

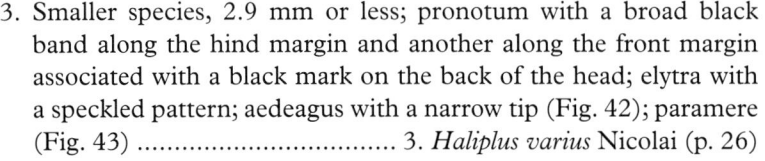

42

43

22

4. Hind margin of the pronotum with distinct, short slots; elytra usually with continuous darker lines along puncture rows, sometimes more blotchy; aedeagus with a hooked tip (Fig. 44); paramere (Fig. 45) .... 1. *Haliplus confinis* Stephens (p. 26)

- Pronotum without any trace of slots; elytra with the dark lines interrupted to produce blotches forming dark transverse bands; aedeagus with a simple tip (Fig. 46); paramere (Fig. 47) ....
.... 2. *Haliplus obliquus* (Fabricius) (p. 26)

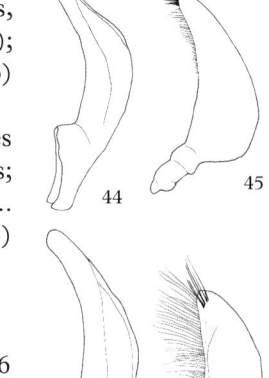

5. Pronotum without any trace of slots; size 2.5-5.0 mm 6

- Short slots extending from one tenth to a third the way forward from the rear of the pronotum; size 2.5-3.2 mm 10

6. Side borders of the prosternal process, as marked by a sharp edge, petering out towards the front of the prosternum (Fig. 48 – tilt the head slightly and illuminate from the side); elytra pale yellow without spots .... 12. *Haliplus flavicollis* Sturm (p. 28)

- Side borders of the prosternal process extending to the front of the prosternum (Fig. 49); elytra dark or pale yellow with or without dark markings 7

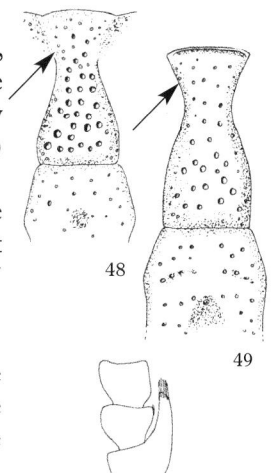

7. Hind edge of the pronotum narrower than the shoulders of the elytra, so that the pronotum and elytra meet with a step, the angle between them being about a right angle, seen from above (Plate 27); male with the basal segment of the mid tarsus shaped like a scoop partly enclosing the next segment of the tarsus (illustrated from the side in Fig. 50) .... 14. *Haliplus laminatus* (Schaller) (p. 29)

- Hind edge of pronotum barely narrower than the front edge of the elytra, the two parts meeting at a wide angle; male with the basal segment of the middle tarsus not scooped out 8

8. Elytra with dark flecks or blotches; the metasternal process usually with a pit at least four times the size of any puncture at the base of the metasternal process (Fig. 51; light from the side, turn the dry beetle in the light, view at x 20 or more) 9

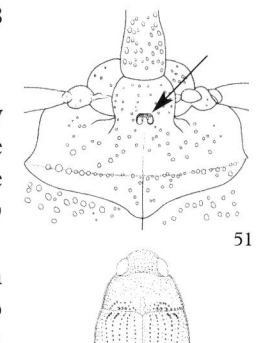

- Elytra uniformly yellow brown, except for the punctures, which are individually dark but not generating any blotches or flecks; no pit at the base of the metasternal process. Body shape (Fig. 52) unlike any other haliplid, with the head and pronotum broad ....
.... 15. *Haliplus mucronatus* Stephens (p. 29)

23

9. Elytra with small dark marks, each covering 1 or 2 rows of punctures; larger species exceeding 3.5 mm in length ....
.... 13. *Haliplus fulvus* (Fabricius) (p. 28)

- Elytra with extensive dark markings each covering 3-4 rows of punctures and often in the form of a cross; smaller species maximally 3.45 mm long ..... 16. *Haliplus variegatus* Sturm (p. 29)

10. Front tarsus with the basal segments enlarged below, and bearing tufts of pale hairs on the underside; basal segment of the middle tarsi swollen compared with the 2nd segment ....
.... Males – *ruficollis* group ... 11

- Front tarsus with all segments symmetrical, without tufts of pale hair beneath; basal segment of middle tarsi not swollen ....
.... Females – *ruficollis* group

The identity of females of this group is best decided by the males with which they are associated.

11. Front tarsus with one claw about ⅔ rds the length of the other and more strongly curved 12

- Front tarsus with the two claws about the same length and curvature (ratio of claw lengths from 1:1 to 4:5) 14

12. Underside of the basal segment of the middle tarsi very clearly concave (depth of concavity much greater than the diameter of a tibial spur); aedeagus with distinctive hook (arrowed in Fig. 53) on dorsal side; right paramere (Fig. 54) with narrow comb and separate tuft .... 8. *Haliplus immaculatus* Gerhardt (p. 27)

- Underside of the basal segment of the middle tarsi almost straight (depth of any concavity less than the diameter of a tibial spur); aedeagus without a distinct hook; right paramere with tuft and comb almost merging 13

53

54

13. Tip of aedeagus finger-shaped (Fig. 55); paramere (Fig. 56) ....
.... 10. *Haliplus ruficollis* (De Geer) (p. 28)

- Tip of aedeagus blunt (Fig. 57); paramere (Fig. 58) ....
.... 11. *Haliplus sibiricus* Motschulsky (p. 28)

14. Underside of the basal segment of the middle tarsi strongly
    concave (Fig. 59); aedeagus similar to that of *ruficollis* but more
    blunt (Fig. 60); right paramere with isolated tuft of bristles at tip
    and main comb in two parts (Fig. 61) ....
    .... 9. *Haliplus lineolatus* Mannerheim (p. 28)

- Underside of the basal segment of the middle tarsi almost straight
    15

15. Pronotal slots straight; aedeagus large with obliquely truncate tip
    (Figs 62 and 64); right paramere with a continuous band of hairs
    along concave edge (Figs 63 and 65); metasternal process with
    depressions on either side, best seen with illumination from the
    side 16

- Slots small and/or curved; aedeagus more rounded at tip (Figs 66
    and 68); right paramere with a lobed inner edge (Figs 67 and 69);
    metasternal process without depressions 17

16. Elytral rows of punctures all strongly pigmented creating six
    continuous but separate lines; aedeagus long and uniformly
    curved along outer edge (Fig. 62); paramere (Fig. 63) ....
    .... 4. *Haliplus apicalis* C.G. Thomson (p. 27)

- Elytral rows of punctures forming discontinuous lines with a
    clear 'window' ⅔ rds of the way from the front formed by the
    third and fourth striae having weakly pigmented punctures (Plate
    19); aedeagus very long and partly concave along upper edge
    (Fig. 64); paramere (Fig. 65) ... 6. *Haliplus furcatus* Seidlitz (p. 27)

17. Larger species, 2.5-2.8 mm; pronotal slots small and weak; elytral pattern a chevron produced by striae being weakly pigmented for 3-8 punctures about a third of the way down (Plate 18); aedeagus long (Fig. 66); right paramere with a large tuft (Fig. 67) ....
.... 5. *Haliplus fluviatilis* Aubé (p. 27)

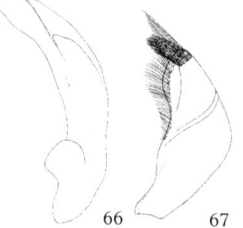

66   67

- Smaller species, 2.2-2.8 mm; slots short and curved; elytral rows with weak pigmentation giving a poorly defined speckled pattern; aedeagus short and with a rounded tip (Fig. 68); right paramere with a narrower tuft (Fig. 69) ... 7. *Haliplus heydeni* Wehncke (p. 27)

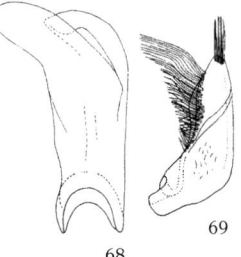

69

68

Subgenus *Haliplidius* Guignot

1. *Haliplus confinis* Stephens Plate 14

Length 3.0-3.5 mm. The fine puncturation of both sexes creating a matt body surface, combined with distinct pronotal slots, should set this species apart. This is the only *Haliplus* with a hook at the tip of the aedeagus. *H. confinis* has both dark and pale forms, the most weakly marked form being common in marl loughs in Ireland. This is a widely distributed species of base-rich waters such as lakes, quarry ponds and grazing fen ditches, usually amongst sparse vegetation and associated with stoneworts on which it feeds. It ranges through the whole of Britain and Ireland, in Scotland on most islands including the Shetlands, and is also known from the Isle of Man and Jersey. Reported throughout the year, peaking in August.

2. *Haliplus obliquus* (Fabricius) Plate 15

Length 3.0-3.5 mm. The pattern of transverse bars on the elytra is distinctive, but pale forms can also occur, particularly in Ireland. The intense puncturing of the body combined with the absence of pronotal slots sets this species apart from all other European species. *H. obliquus* is mainly associated with base-rich habitats supporting stoneworts, on which it feeds. Common in the east of England but more scattered in the west and in the north, confined in Wales to the coast, in particular on Anglesey, and common on the karst of Ireland, also on the Isle of Man. In Scotland it is rare even in the lowlands, but is common on the north-eastern tip of Caithness, being also recorded from Islay, Lismore, Orkney and Tiree. Reported throughout the year, peaking in June and August.

3. *Haliplus varius* Nicolai Plate 16

Length 2.6-2.9 mm. *H. varius* looks like a small *H. obliquus*, but the markings are more speckled, with strong black marks front and rear on the pronotum. There are also very small black slots hidden in the hind black mark of the pronotum. The extremity of the aedeagus, i.e. that part beyond the opening of the sperm duct, is shorter than in *H. obliquus*. This is a rare English species, known only from two old pond systems in Sussex, and only from one of those recently. This species has been found in September and October.

Subgenus *Haliplus* Latreille

4. *Haliplus apicalis* C.G. Thomson Plate 17

Length 2.5-3.0 mm. The strong lines on the elytra are unusual amongst the entire *H. ruficollis* group but the male genitalia must be examined in order to be certain as some *H. ruficollis* can also be striped in this way. A predominantly eastern seaboard species both in Britain and Ireland found in brackish waters such as coastal lagoons, puddles and drainage ditches, also occasionally inland. The northernmost confirmed records are for the Solway and Northumberland. This species has been reported in all months except December, and shows a very strong peak in May and a lesser one in August.

5. *Haliplus fluviatilis* Aubé Plate 18

Length 2.5-2.8 mm. *H. fluviatilis* is pale yellow and, if present, the chevron mark on the elytra is distinctive. However, some specimens are spotted like other species. This species is typical of large rivers nearing their estuaries but also occurs in man-made pools and larger drainage ditches with plenty of exposed substratum in lowlands in Britain and Ireland. There is only one old record north of a line from the Clyde to the Forth. Reported throughout the year, peaking in August.

6. *Haliplus furcatus* Seidlitz Plate 19

Length 2.4-3.0 mm. The pale "windows" on the rear half of the elytra show up under the hand lens in the field, and the large aedeagus is distinctive. The habitat is relict rich fen pools and similar habitats in Jersey, Norfolk, Oxfordshire and Somerset, and the species can appear fleetingly in numbers. This species has been found in March, April, June and October, with most records for April.

7. *Haliplus heydeni* Wehncke Plate 20

Length 2.2-2.8 mm. This small species is often found with *H. ruficollis*. Additional characters not used in the keys, but useful where these species coexist, are that the females of *heydeni* have micropunctures at most around the hind edges of the elytra whereas such punctures cover a variable part of the elytra of *H. ruficollis*; also the males of *H. ruficollis* have unequal fore claws, whereas these are about the same length in *H. heydeni*. A southern species of small ponds, often in part shade, confined to Wales and England north to Westmorland. This species has been reported in all months except December, with peaks in April and September.

8. *Haliplus immaculatus* Gerhardt Plate 21

Length 2.4-3.1 mm. This a pale species with an elytral pattern mainly based on four continuous stripes on the four inner rows of punctures. The basal segment of the mid tarsus is curved but not as strongly as in *H. lineolatus*. The long and strongly curved aedeagus with its backwardly pointed hook can often be seen without dissecting the genitalia completely. Mainly associated with man-made stagnant water habitats in the lowlands, even in polluted sites such as motorway balancing lagoons. This species extends north to Fife with scattered old records to East Sutherland and its only known Scottish island, Coll. Known from Guernsey and Jersey. Reported in all months of the year, peaking in June and August.

9. *Haliplus lineolatus* **Mannerheim** Plate 22

Length 2.5-3.2 mm. This is a small species, with the most distinctive feature being the strongly sinuate basal segment to the male mid tarsus, viewed from the side. The aedeagus is similar to that of *H. ruficollis* and the tarsus should always be checked. *H. lineolatus* is usually found in larger ponds, lakes and slow sections of rivers. It is known to feed on Hydrozoa but will also feed on algae. Mainly a northern species, barely reaching the southern coast of England and Wales, apparently lost from much of the south of Ireland, reaching Bute, Lismore, Mull and South Uist. Reported in all months of the year, peaking in June and August.

10. *Haliplus ruficollis* **(De Geer)** Plate 23

Length 2.5-3.0 mm. This is a nondescript species, the most distinctive feature being the narrowed tip of the aedeagus. It is the commonest species in Ireland, and also in Britain apart from Shetland and northern mountainous areas. It is known from Guernsey and Jersey. Reported in all months of the year, peaking in June and August.

11. *Haliplus sibiricus* **Motschulsky** Plate 24

Length 2.5-3.3 mm. There are no reliable external characters for separating this species from *H. ruficollis*, but the blunt tip of the aedeagus, which can often be seen without a full dissection, should be enough. A widespread species commonest in northern England, associated with many man-made habitats in still and slow-running water, and also in smaller natural lakes and rivers in the north, usually in lowlands. Scattered in Scotland, recorded from Islay, Lewis, Lismore, and the Orkneys. It is known from the Isle of Man and there is an old record for Jersey. Reported in all months of the year, peaking in August. Formerly known as *wehnckei*.

Subgenus *Liaphlus* Guignot

12. *Haliplus flavicollis* **Sturm** Plate 25

Length 3.5-4.0 mm. This is one of our three large species, normally distinguished from *H. fulvus* by its uniform yellow colour. Specimens with weak blotches can be separated from *H. fulvus* by the prosternal process character, and also by the setiferous striole running over more than half the length of the hind tibia. Both are difficult features to see with conviction. *H. flavicollis* is widely distributed in lowland base-rich waters such as lakes, reservoirs, brick pits, and slow drains over exposed substratum with thin vegetation, and including fish-stocked sites. Absent from much of northern Scotland, reaching only the islands of Barra, Rum and South Uist. Known from Jersey. Reported in all months of the year, peaking in June and August.

13. *Haliplus fulvus* **(Fabricius)** Plate 26

Length 3.6-4.2 mm. This is our only large spotted species. The setiferous striole supposedly runs for about a third of the length of the hind tibia but the best character if in doubt about the spots are the sharp edges of the prosternum running up to the head. *H. fulvus* occurs in the same places as *H. flavicollis*, but extends into more acid and upland waters. Some specimens, mainly from the Scottish Highlands, have the spots reduced. Known from Jersey. Reported in all months of the year, peaking in June and August.

14. *Haliplus laminatus* **(Schaller)** Plate 27

Length 2.5-3.0 mm. The pale greyish yellow body colour is distinctive, as are the strong shoulders of the elytra standing out behind the pronotum. A unique feature is provided by the basal segment of the male mid tarsus, which is scoop-shaped and encloses the next segment. *H. laminatus* can be mistaken for members of the subgenus *Haliplus*, i.e. the *Haliplus ruficollis* group, but it lacks the slots on the pronotum. A southern English species of man-made habitats, particularly canals and ponds, plus slow sections of rivers, also in the River Wye in Monmouthshire. Reported in all months of the year except February, peaking in May and September.

15. *Haliplus mucronatus* **Stephens** Plate 28

Length 4.0-5.0 mm. This is our largest species at up to 5 mm and is uniformly an unpatterned yellow or yellowish brown, so the only species with which it is likely to be confused should be *H. flavicollis*. However, *H. mucronatus* has an unusually wide head and pronotum (Fig. 52), though these are difficult to measure usefully. The smooth prosternum and metasternum also set this species apart from the other large species. The habitat is typically on clay, including natural subsidence ponds but mainly man-made stagnant water habitats. The main distribution of this species lies from northern East Anglia to Humberside, with outlying populations on the grazing fens of the Weald and around the coast of the Bristol Channel into south Wales. Reported in all months of the year except January, peaking in April and September.

16. *Haliplus variegatus* **Sturm** Plate 29

Length 2.5-3.5 mm. *H. variegatus* looks like a small *H. fulvus*, though with the dark marks usually coalescing into a strong pattern with much darkening on the rear two-thirds of the centre around the elytral suture. This is an uncommon species of stagnant fens, either on soft peat or on clay, and always in association with its food, stoneworts. It is largely confined to relict lowland habitat with clean water such as in the New Forest, the Norfolk Broads, and the Lizard and also in coastal grazing fens. Reported in all months of the year except January, peaking in April and September.

Subgenus *Neohaliplus* Netolitzsky

17. *Haliplus lineatocollis* **(Marsham)** Plate 30

Length 2.6-3.5 mm. This species should not be mistaken for any other, with its dark pronotal markings and the long slots. Also the insect often appears tri-coloured, with the head reddish, the pronotum a stronger yellow and the elytra a greyish yellow. The underside is darkened. A widespread species and usually the commonest haliplid in running water in Britain and Ireland. Common on Guernsey and Jersey. Reported throughout the year, peaking in August.

3. *PELTODYTES* Régimbart

1. *Peltodytes caesus* (Duftschmid) Plate 31

Length 3.5-4.0 mm. This is a broad species with a superficial resemblance to *Haliplus lineatocollis* in markings: the larger metacoxal plates and the long ultimate segment to each maxillary palp should prove enough to separate it from *Haliplus* spp. Confined to lowland rich fen pools and ditches from the Welsh and English fens surrounding the Bristol Channel and from the Isle of Wight to Norfolk. It is frequent inland in the lowlands to the west of London as far as Oxfordshire and North Hampshire. There is an old record from Jersey. Reported throughout the year except December, peaking in April and September.

Family NOTERIDAE C.G. Thomson – Burrowing diving beetles

1. *NOTERUS* Clairville

The two species are grouped with the Dytiscidae as diving beetles, characterised by the synchronised movements of the hind legs to produce a strong diving thrust. The males of *Noterus* have segments 4-11 of the antennae expanded, those of the female having smaller expansions on segments 7-9. The elytra have scattered large pits in both sexes. The larvae attach themselves to the aerenchymatous tissue of aquatic plants such as sweet grasses (*Glyceria* spp.) and bog bean (*Menyanthes trifoliata* L.) in order to renew their air supply, and the pupae are similarly attached. Thus the adults are mainly associated with vegetated edges or rafts of vegetation.

Key 5. The species of *Noterus*

1. Larger species, length 4.0-5.0 mm; prosternum with a ridge in the mid-line, running from a minute point on the front margin back along the process; end of the process with a narrow neck (Fig. 70) .... 1. *Noterus clavicornis* (De Geer)

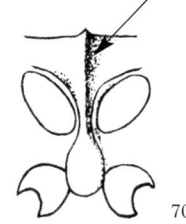

- Smaller species 3.5-4.0 mm; prosternum without a mid-line ridge; front margin of the prosternum smooth; process with a wide neck (Fig. 71) .... 2. *Noterus crassicornis* (O.F. Müller)

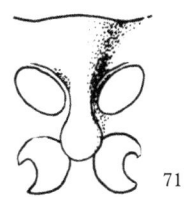

1. *Noterus clavicornis* (De Geer) The Larger Noterus Plate 32

Length 4.0-5.0 mm. Common in permanent, base-rich, lowland ponds in England, Ireland and Wales, spreading in Scotland into the Central Belt, reaching Lundy and the Isle of Man, but so far known from only one Scottish island, Bute. It is found on Alderney, Guernsey and Jersey. The larger species was incorrectly known previously as *N. capricornis*, also as *sparsus* (Marsham), a rare case where the proposed common name is potentially less confusing than the accepted Latin name. Reported throughout the year, peaking in May and September.

2. *Noterus crassicornis* (O.F. Müller) The Smaller Noterus Plate 33

Length 3.5-4.0 mm. This species is found in permanent, base-rich lakes, ponds and grazing level drainage ditches. It is scarcer than *N. clavicornis* with a patchy distribution, common in much of Ireland, and frequent in East Anglia, the Trent Valley, the Vale of York, the Cheshire Plain, and Anglesey, otherwise in isolated places from the Scillies to two sites in Scotland, the water bodies of the former area of Carlingwark Loch, Castle Douglas, and Lindores Loch in Fife. A synonym for this species is *capricornis*, though this has been more often applied, incorrectly, to the larger species. An alternative name proposed to overcome this confusion, *minor* Balfour-Browne, is also a synonym. Reported throughout the year, peaking in May and September.

Family PAELOBIIDAE Erichson

1. *HYGROBIA* Latreille

1. *Hygrobia hermanni* (Fabricius) The Squeak Beetle Plate 34

Length 8.5-10 mm. This distinctive beetle, which usually announces its presence when caught, is the sole Palaearctic representative of this family, another species of *Hygrobia* being known from China and four from Australia. It is confined to still water, usually over mud in ponds and ditches where it feeds on chironomid larvae and oligochaete worms. *H. hermanni* is frequent across lowland England and Wales, and the southern half of Ireland being known even in mountain pools; it reaches County Durham in the north-east and has been found since 1999 in Westmorland, from Cumberland since 2006, and for the first time on the Isle of Man in 2009. A single specimen was found in Kirkcudbrightshire in September 2006. It is known from Alderney, Guernsey and Jersey. *H. hermanni* has been reported in all months, peaking in May/June and September; the larva, easily distinguished from other water beetle larvae by its three-pronged tail, being found from June to September, peaking in June.

Family DYTISCIDAE Diving beetles

The main group of diving beetles is easily recognised in the field by a fast, smooth swimming action, the hind legs thrusting in unison. Dytiscidae share the simple, filamentous, 11-segmented antennae with Haliplidae and Paelobiidae among the beetles capable of swimming but the latter two families swim by alternating movements of the legs, and thus waver in their path. The Noteridae, which swim like Dytiscidae, can be differentiated by their thickened antennae.

Dytiscidae have a major modification to the thorax, the second segment (mesothorax) being scarcely visible beneath the underside of the greatly expanded metathorax, the metasternum, meeting the backwardly pointing process of the first thoracic segment (prosternal process). The largest plates on the underside of the skeleton are derived from the metacoxae, which provide many useful features at species level in addition to distinguishing the group as a whole.

Body length ranges from about 1 mm to over 30 (Figure 72, p. 32) but the streamlined structure is similar whatever the size. It is this smooth outline that renders some Dytiscidae so difficult to identify with certainty. In addition to size limits within species Dytiscidae provide an example of "self-organised similarity" (Scheffer & Nes, 2006). The three most species-rich subfamilies have a stepped sequence in body size – Hydroporinae 1.9-5.5 mm, Agabinae 5.8-14.5 mm, Dytiscinae 14-37 mm – echoing a similar phenomenon in the Carabidae. Thus size can also be used to recognise the main groups and genera.

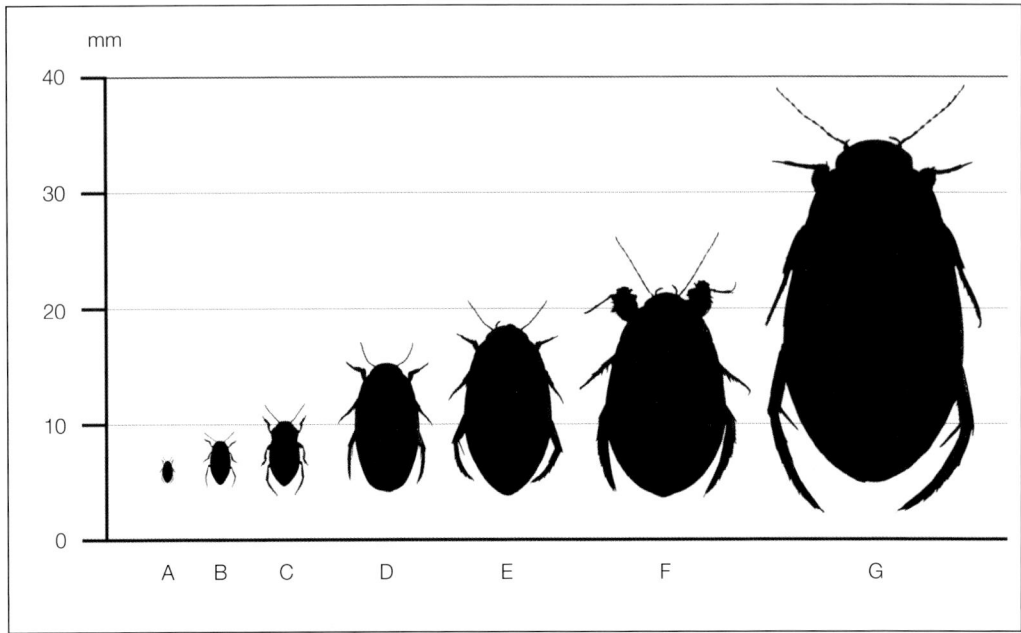

Figure 72. Some examples illustrating size extremes in Dytiscidae: A – *Hydroglyphus geminus*; B – *Hydroporus palustris*; C – *Stictotarsus duodecimpustulatus*; D – *Agabus bipustulatus*; E – *Ilybius ater*; F – *Acilius sulcatus*; G – *Dytiscus marginalis*

Whilst it might seem desirable to key out the family systematically, i.e. firstly to subfamily or tribe, then genus, then species, the rather limited British and Irish faunas make it possible to tackle some poorly represented genera on what might seem quite trivial characters. The first key takes out such poorly represented genera and leads onto sixteen other keys including two to the major problem groups, *Hydroporus* and *Agabus + Ilybius*. Some subfamilies, tribes, genera and subgenera are still controversial with their precise definition in flux varying dependent on analyses of adult morphology, bristle disposition in larvae and DNA sequence data, plus, of course, varying interpretations of such data. Several species of Dytiscidae have dimorphic females, with one resembling the male and the other with grooves or microreticulation that are thought to test the copulatory fitness of males. The males associated with such "resistant" females may exhibit a morphological response, usually an increase in the number of sucker hairs on the tarsi. In a few cases, e.g. *Hydroporus memnonius* and *Agabus uliginosus*, the distributions of the two forms differ considerably and give cause for considering them to be functionally separate, demanding recognition as distinct entities ecologically and biogeographically.

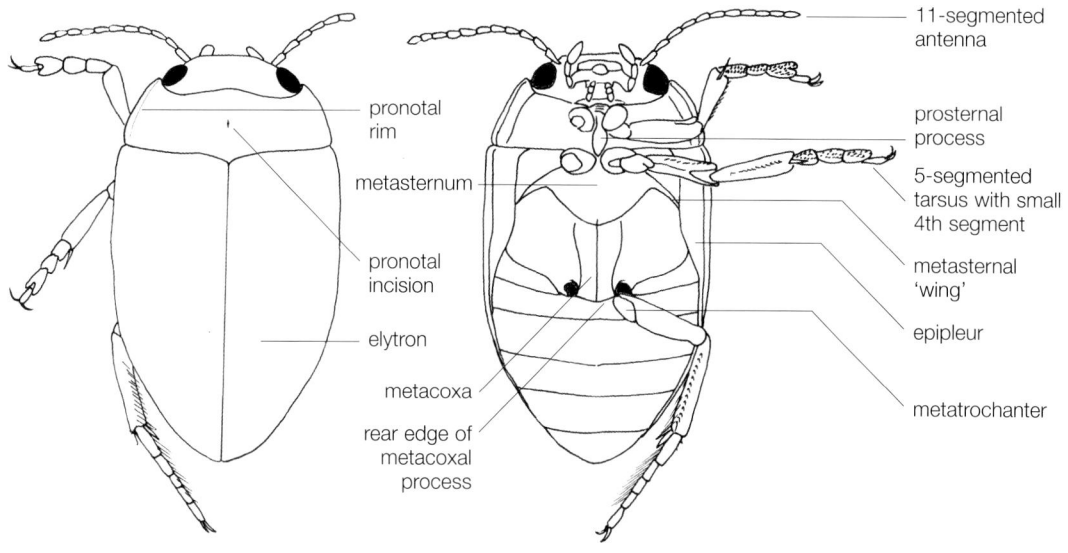

Figures 73 and 74. Upper and under sides of a diving beetle

Key 6. The genera of Dytiscidae

1. Small triangular plate (scutellum) visible at the junction of the pronotum with the elytra (Fig. 75); length 6.0 mm or more 2

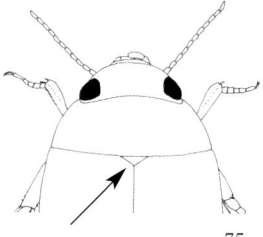

75

- Scutellum absent, but pronotum may extend backwards into a point in the mid-line (Fig. 76); length 6.0 mm or less 11

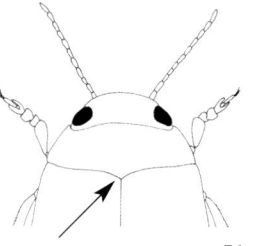

76

2. Length 22 mm or more 3

    the larger Dytiscinae

- Length 18 mm or less 4

3. Hind tibiae longer than wide and longer than the spurs (Fig. 77); paired hind claws; females usually with grooved elytra (Plates 83, 88 and 90); length 22-39 mm ....
.... 10. *Dytiscus* (6 spp.) Key 11 (p. 65)
Dytiscinae

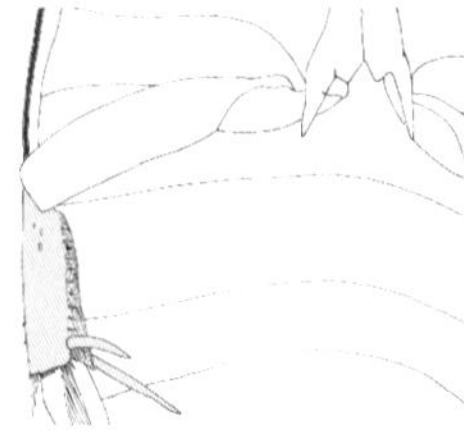

77

- Hind tibiae barely longer than wide with one tibial spur massive compared to the other one and both almost as long as the whole tibia (Fig. 78); single hind claws; all females without grooves; length 29-37 mm .... 9. *Cybister* (1 sp.) (p. 64)
Dytiscinae

Tibia and spurs are grey in Figs 77 and 78. The epipleur is black.

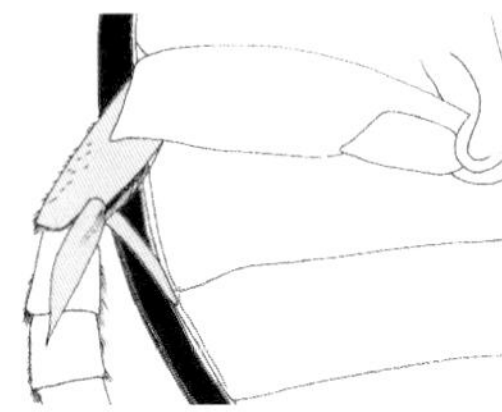

78

4. Eyes straight-edged at the front (Fig. 79); male front tarsi almost circular with hairs on the undersides carrying a mixture of large and small sucker cups (Fig. 80) 5

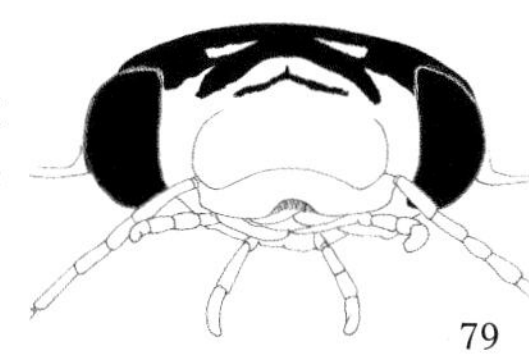

79

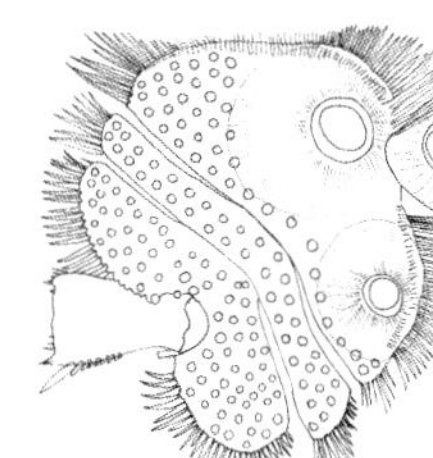

80

- Eyes distinctly indented above where the antennae are inserted (Fig. 81); male tarsi wider than in the female but not rounded and with hairs bearing only small sucker cups 7

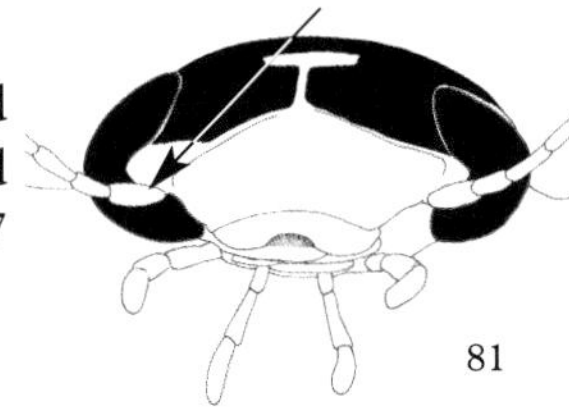

81

5. Upper side black with yellow margins (Plates 91 and 92); length 12-15 mm .... 11. *Hydaticus* (2 spp.) Key 12 (p. 68)
Dytiscinae

- Upper side greyish brown or yellow with a black network (Plates 76-78); length 12-18 mm 6

6. Underside mainly black; upper side greyish brown with some flecks (Plates 73-75); females with hair-filled grooves on the elytra; length 14-18 mm .... 7. ***Acilius*** (2 spp.) Key 9 (p. 62)
   Dytiscinae

\- Underside mainly orange brown; upper side ground colour yellow with black meshes on elytra (Plates 76-78); females smooth like the males; length 12-16 mm ....
   .... 8. ***Graphoderus*** (3 spp.) Key 10 (p. 63)
   Dytiscinae

7. Large grey and brown species (length 15-17 mm) (Plate 65); elytra with transverse striae as the stronger feature of a double reticulation (Fig. 82) .... 4. ***Colymbetes*** (1 sp.) (p. 58)
   Colymbetinae

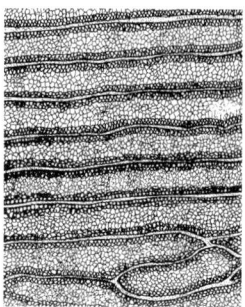

82

\- If large (length 15 mm) then black, otherwise less than 13 mm long; any elongate part of the elytral reticulation running front to rear 8

8. A reddish, elongate species with distinctive elytral sculpture, many longitudinal cuts on the reticulated surface (Fig. 83); lobes of the hind coxal process diverge widely (Fig. 84); lines on the hind coxal processes almost touching in the mid-line then turning sharply through 90° to run onto the lobes; length 6.3-7.9 mm ....
   .... 6. ***Liopterus*** (1 sp.) (p. 61)
   Copelatinae

83

\- Not as above, either smaller or if reddish then not so narrow, many types of elytral sculpture but none as illustrated for the species above; hind coxal lobes diverge only slightly; hind coxal lines always widely separate never turning so abruptly onto the lobes ....
   9

84

9. Epipleurs (view underside of elytron – left with the elytron black, Fig. 85) remaining broad towards the rear; prosternal process with flared edges (Fig. 87); a single species with black elytra with yellow flecks, and the pronotum often with a broad white band across its front .... 3. ***Platambus*** (1 sp.) (p. 57)
Agabinae

- Epipleurs more narrow in the posterior half, sometimes with a distinct contraction at about the halfway point (Fig. 86); prosternal process not so flared; not patterned as in *Platambus maculatus* .....
.... 10
the rest of the Agabinae

10. The outside rear corners of the undersides of the hind femora each with a line of bristles, sometimes as few as three (here illustrated for *Agabus nebulosus*), arranged in the form of a short comb (Fig. 88) .... 1. ***Agabus*** (18 pp.) and 2. ***Ilybius*** (10 spp.) Key 7 (p. 42)
Agabinae

These two genera cannot be separated consistently without examination of the male genitalia: they are identified to species in a single key.

- The outside rear corners of the undersides of the hind femora each without bristles or, if a few are present, weak and not arranged as a comb (Fig. 89) ... 5. ***Rhantus*** (6 spp.) Key 8 (p. 58)
Colymbetinae

11. Hind tarsi with broad segments with long lobes underneath (Fig. 90); front and mid tarsi with 5 simple segments (Fig. 91); streamlined species, able to escape capture by "skipping" when out of the water; body length 3.5-5.1 mm ....
.... 29. ***Laccophilus*** (3 spp.) Key 22 (p. 107)
Laccophilinae

- Segments of the hind tarsi narrow and simple (Fig. 92); front and mid tarsi 5-segmented but apparently with 4 segments, the deep lobes of the 3rd surrounding a diminutive 4th segment (Fig. 93); body sometimes streamlined but often with a slightly irregular body outline, more likely to crawl than skip when out of water; body length 1.5-5.5 mm 12
Hydroporinae

12. Each elytron with a small tooth projecting from the margin near the tip (Fig. 94) .... 17. ***Nebrioporus*** (3 taxa) Key 16 (p. 91) **Hydroporinae**

94

- Elytra without teeth at rear 13

The females of *Oreodytes alpinus* have the elytra flared at the tip, rather like large teeth (see Key 17 if in doubt).

13. Three weak ridges (left side in Fig. 95) overlying the five more or less intact black stripes (right side in Fig. 95) on each elytron .... 17. ***Nebrioporus*** (1 sp., ***canaliculatus***) (p. 92) **Hydroporinae**

95

- Elytra without ridges, usually smooth, sometimes with incisions 14

14. Pronotum with two longitudinal furrows placed ⅔ rds of the way from the mid-line to the lateral margin and matched by a pair of furrows on the elytra (Fig. 96); length 1.5-2.0 mm 15

Longitudinal furrows are marked by a sharp vertical edge, visible as a dark line when illuminated from the front or side. Distinguish such an edge from a line of pigment or one of a number of similar rows of punctures.

96

- Pronotum without furrows in that position; mostly larger than 2 mm 16

15. The grooves running alongside the suture (sutural striae) reaching to behind a dark central patch, and usually to the extremity (Fig. 97); elytral pattern distinctive, being a large blotch with variable incomplete longitudinal stripes (Fig. 97) .... 13. ***Hydroglyphus*** (1 sp.) (p. 69) **Hydroporinae**

97

- Sutural striae fading no more than ⅔ rds of the way down the elytra (Fig. 96); elytra either dark or with transverse bands coalescing over a yellow background (Fig. 98) .... 12. ***Bidessus*** (2 spp.) Key 13 (p. 69) **Hydroporinae**

98

16. Elytra drawn out into a small but distinct point at the rear (Fig. 99); broadly oval reddish brown species; prosternal process unusual, having a blunt rear edge (Fig. 100); length 2.2-2.5 mm .... 25. *Hydrovatus* (2 spp.) Key 19 (p. 99)
Hydroporinae

- Elytra not drawn out into a distinct point; variously shaped and coloured, sometimes broad oval species; prosternal process more elongate, raised and usually spear-shaped; mainly larger than 2.5 mm 17

99

100

17. Pronotum on each side with a longitudinal furrow (Fig. 101) defined as a sharply edged slot in dark species, weaker in pale ones 18

View the beetle with light from the side making sure that the grooves are not obscured by dirt or matted hair. The slots can be picked out by light shining on the edge though the shadow of the edge may be more visible.

- Pronotum without longitudinal furrows 21

101

18. Colour predominantly black with yellow markings; hind coxal process with rounded lobes (Fig. 102); length 2.0-3.0 mm 19

- Colour predominantly yellow with black markings; hind coxal processes diverging and only slightly rounded; length 3.0 mm or more 20

102

19. Elytra with transverse pale wavy markings (Fig. 103); length 2.9-3.1 mm .... 22. *Stictonectes* (1 sp.) (p. 96)
Hydroporinae

- Each elytron with at least one longitudinal pale stripe running from the pronotum to near the tip; length not more than 2.8 mm .... 15. *Graptodytes* (4 spp.) Key 14 (p. 70)
Hydroporinae

103

20. Upper side microreticulate and without hair ....
.... 18. ***Oreodytes*** (4 spp.) Key 17 (p. 93)
Hydroporinae

\-   Upper side shining without microreticulation beneath a covering
of fine hair .... 29. ***Scarodytes*** (1 sp.) (p. 96)
Hydroporinae

Scarodytes is keyed out twice to take into account the difficulty of detecting the
lateral furrows on the pronotum in some individuals.

21. The outer corner of the underside of the elytron with a distinct
ridge dividing the epipleur into two parts (Fig. 104); length only
about twice to 2¼ times maximum depth viewed from the side
(Fig. 105) 22

104

105

\-   No dividing ridge on the epipleur; length about 2½ to 3 times
maximum depth (Fig. 106) 23

106

22. Globular beetles, occasionally with a weak mottling on the elytra,
rarely (Channel Isles only) with a distinctive elytral pattern of
bars and partial stripes resembling bats with their wings partly
outstretched – see Plate 157); hind claws unequal, the shorter
one (arrowed Fig. 107) being barely wider than the spines; length
4.5-5.0 mm .... 27. ***Hyphydrus*** (2 spp.) Key 21 (p. 106)
Hydroporinae

107

\-   Elytra with distinctive colour patterns; either longitudinal lines or
blotches; hind tarsal claws equal in length; 2.0-4.0 mm ....
.... 26. ***Hygrotus*** (9 spp.) Key 20 (p. 100)
Hydroporinae

23. Neck of prosternal process without a distinct step (Fig. 108),
often with a covering of thick hairs (Fig. 109) 24

109

108

\-   Neck of prosternal process with a sharply raised edge defining a
distinct step (Figs 110 and 111) 27

110

111

24. Body shape more squat (Plate 96); red or reddish brown without stripes or sharply defined spots, sometimes with a pale front margin to the elytra; antennae pale throughout; pronotum and elytra with many small punctures and scattered pits 4-5 times the diameter of the small ones, plus a few hair-bearing pits in lines, with a lattice-like structure visible only at x 80 and obscured by the densely packed small punctures (Fig. 112) ....
.... 14. *Deronectes* (1 sp.) (p. 70)
Hydroporinae

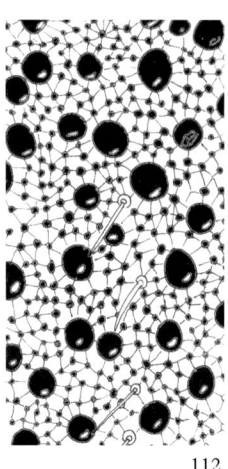

112

- Body shape more elongate; striped or spotted; antennae darkened or pale throughout; upper side finely punctured, all punctures being about the same size 25

25. Black, usually with a reddish orange head and similarly coloured sides to the pronotum and elytral blotches; prosternal process a long oval (Fig. 113), slightly domed; male inner fore claw lobed .... 24. *Suphrodytes* (2 spp.) Key 18 (p. 97)
Hydroporinae

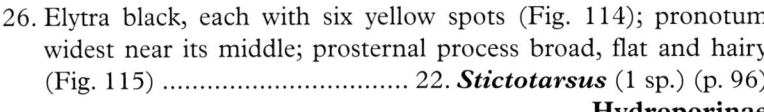

113

- Mainly orange-yellow with elytral pattern of spots or stripes; prosternal process either broad, flat and almost hexagon-shaped or narrow and keeled; male inner fore claw not lobed 26

26. Elytra black, each with six yellow spots (Fig. 114); pronotum widest near its middle; prosternal process broad, flat and hairy (Fig. 115) .... 22. *Stictotarsus* (1 sp.) (p. 96)
Hydroporinae

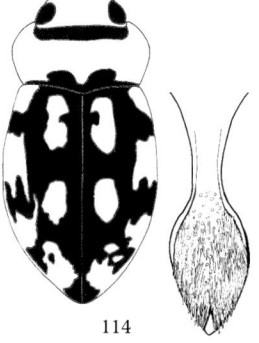

114

115

- Elytra yellow, each with five coalescing black stripes; pronotum widest at its base (Fig. 116); prosternal process narrow, spear-shaped with a crest of curved hairs covering the central ridge (Fig. 117) .... 23. *Boreonectes* (1 sp.) (p. 97)
Hydroporinae

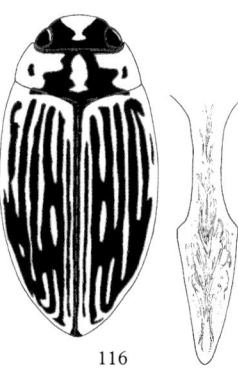

116

117

27. Elytra with roughly equal numbers of small and large punctures on a surface visibly microreticulate at × 40 (Fig. 118), pronotum with only the fine punctures; length 4.5-5.0 mm ....
.... 28. ***Laccornis*** (1 sp.) (p. 106)
Hydroporinae

Body shape distinctive with elongate elytra (Plate 159); two-tone colouring distinctive with pronotum darker than brown elytra; most antennal segments darkened.

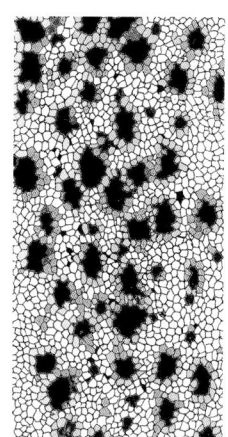

118

- Middle part of elytra with scattered punctures all of about the same size, though there may also be lines of larger punctures; length mostly less than 4 mm, a few species reaching 5.3 mm in length 28

28. Metacoxal processes not meeting posteriorly at the midpoint, the process edges being either rounded (Fig. 119) or straight and pointed inwards; elytra with long stripes; length 3.0-4.3 mm ... 29

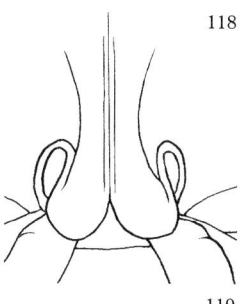

119

- Metacoxal processes meeting in the midline at the rear, the process edges being sinuous (Fig. 120) or straight (Fig. 121); elytra not striped; length 1.9-5.3 mm ....
...................................16. ***Hydroporus*** (28 spp) Key 15 (p. 72)
Hydroporinae

The beetle should be manipulated along with the light source to achieve the best chance of viewing the edges of these processes.

120

121

29. Pronotum, elytra and underside reddish yellow, elytra with obscure striped pattern often visible only under the microscope (Fig. 122); 3.0-3.5 mm .... 19. ***Porhydrus*** (1 sp.) (p. 95)
Hydroporinae

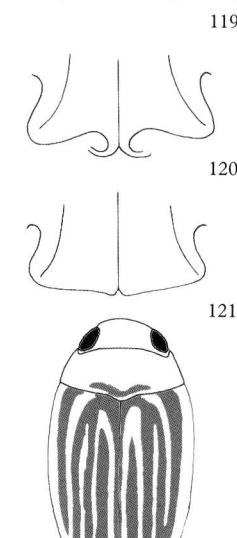

122

- Pronotum yellow with two central dark spots (Fig. 123); elytra dirty yellow with distinct dark stripes visible to the unaided eye; underside black; length 3.8-4.3 mm ....
.... 20. ***Scarodytes*** (1 sp.) (p. 96)
Hydroporinae

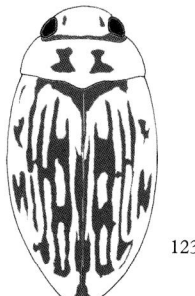

123

Subfamily AGABINAE Thomson

1. *AGABUS* and 2. *ILYBIUS*

Key 7. The species of *Agabus* and *Ilybius*

In these genera it is only the males that possess sucker hairs on the fore and mid tarsi: these tarsi are also appreciably wider than in the females. This key depends to a considerable extent on dissection of the male genitalia, reinforced where possible by some male secondary sexual characters. This underlines the danger of trying to identify isolated females of some species. General appearance, as to be seen in the plates, is useful if only to demonstrate how similar some species are. Body length is also important. The ability to see microsculpture at high magnification (x 60 and upwards) is required. Colour can be affected by immersion in preservative – the insects should be dry in order to see whether they have a metallic finish, and to detect the two-tone appearance of some species with lighter elytra contrasting with the dark pronotum.

1. A dull black species; 12.5-14.5 mm long ....
   .... 2. *Ilybius ater* (De Geer) (p. 55)

- Species not more than 12.5 mm long, the largest having a slight
  metallic tinge 2

2. Elytra with a distinct double reticulation throughout, a meshwork
   within a meshwork 3

- Elytra may be reticulate or shining or doubly reticulate only in
  small areas 6

3. Pronotum black with orange margins, elytra brown with front
   outer corners orange (Plate 37); the larger meshwork of the elytra
   scarcely longer than wide (Fig. 124); male – aedeagus with a
   double point (Fig. 125); length 7.7-8.9 mm ....
   .... 3. *Agabus sturmii* (Gyllenhal) (p. 50)

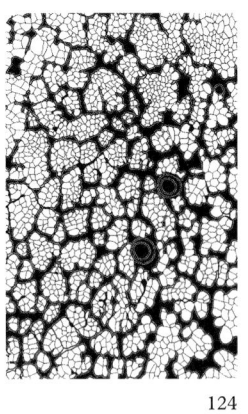

124

125

- Black, with at most a pair of spots on the head and margins of
  pronotum dark red; either the larger meshwork of the elytra
  elongate or, if not longer than wide, then the intersections of the
  larger meshes marked by pits; male aedeagus with a single point
  4

Beware of teneral specimens that may be paler than usual; teneral specimens have
a soft cuticle, easily dented or torn.

4. Body with a continuously rounded outline, widest at about the midway point; primary reticulation elongate (Fig. 126); length 9.5-11.6 mm; aedeagus pointed (Fig. 127) ....
.... 9. *Agabus bipustulatus* (Linnaeus) (p. 52)

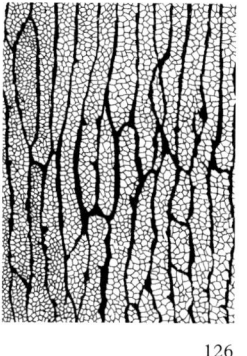

126

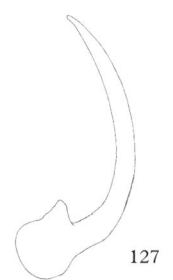

127

- Body almost parallel-sided where the elytral edge is straight; primary reticulation elongate or isodiametric; not more than 8 mm long; aedeagus with a blunt tip (Figs 129 and 131) 5

5. Larger meshes of elytral reticulation elongate with intersections simple (Fig. 128); aedeagus broad with the underside of the apex finely saw-toothed (Fig. 129); 7.2-7.9 mm ....
.... 17. *Agabus striolatus* (Gyllenhal) (p. 54)

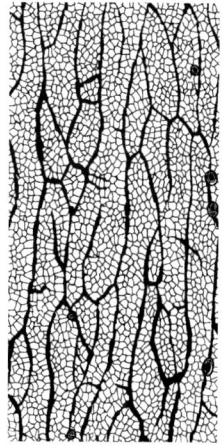

129

128

- Larger meshes of elytral reticulation about as wide as long, and with many of the intersections marked by pits (Fig. 130); aedeagus narrow with underside smooth (Fig. 131); length 6.4-7.3 mm ......
.... 10. *Ilybius wasastjernae* (C.R. Sahlberg) (p. 57)

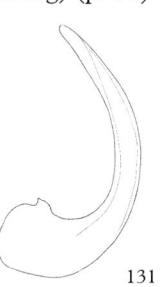

131

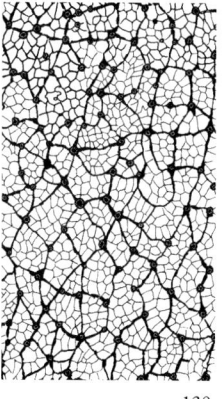

130

6. Pronotum and elytra a pale straw colour with some mottling and spotting 7

- Pronotum and elytra mostly black or dark brown 8

7. Pronotum with two dark spots and elytra with well defined, small, irregularly shaped spots ...... 15. *Agabus nebulosus* (Forster) (p. 54)

- Pronotum without spots; elytra with vague, cloud-like markings .... 11. *Agabus conspersus* (Marsham) (p. 53)

Very rarely, *A. nebulosus* lacks the pronotal spots but its general colouring and interconnected dark spots on the elytra should suggest that it is this species rather than *conspersus*: the claw-bearing segment of the fore leg of male *nebulosus* has hairs confined to the outer two-thirds, whereas slightly finer hairs run along almost all of the same segment in male *conspersus*.

8. Antennae with last segment all pale like the rest 9

- Antennae with the last segment at least and usually the 5th to 10th partly or wholly darkened, also the extremities of the palps 17

This is a good character even though it cuts across genera and subgenera. For specimens with the last segment broken off the 5th segment and the rest outwards are often also dark in species with the last segment dark, and the extremities of the palps are also usually dark in such species.

9. Hind claws unequal in length (Fig. 132) 10

- Hind claws similar in length 15

10. Metasternal extension, the "wing", very narrow (Fig. 133); body outline broadest behind the middle; upper side bronze and underside reddish .... 4. *Ilybius fenestratus* (Fabricius) (p. 56)

All diving beetles can produce corticosteroid substances from prothoracic glands. *I. fenestratus* is particularly pungent, producing testosterone.

- Metasternal "wing" tapering more evenly to the edge (Fig. 134); body outline broadest at the midway point; upper side bronze or black with underside dark red or black 11

11. Viewed from the side a yellow streak runs along the margins from just behind the eye to ⅔ rds of the way down the elytra encompassing the paler elytral spots; otherwise bronze .... 5. *Ilybius fuliginosus* (Fabricius) (p. 56)

- No yellow streak, at most a few spots and slight lightening of the edges of the pronotum and elytra; black or bronze insects 12

12. Upper side devoid of a metallic sheen 13

- Upper side with a metallic sheen 14

13. Length 10.5-12.2 mm; aedeagus slightly but distinctly turned down at its tip, this feature sometimes visible without complete dissection (Fig. 135); inner fore claw of male with a tooth .... 8. *Ilybius quadriguttatus* (Lacordaire) (p. 57)

- Length 8.7-10.0 mm; aedeagus smoothly narrowing to a tip resembling the prow of a boat (Fig. 136); fore claws not toothed .... 6. *Ilybius guttiger* (Gyllenhal) (p. 56)

14. Length 8.5-9.8 mm; black save for the faint metallic sheen; aedeagus evenly tapering to a narrow blunt point (Fig. 137) .... 1. *Ilybius aenescens* C.G. Thomson (p. 55)

This species can also be differentiated from the similarly sized *I. guttiger* by its pungent smell.

- Length 10.0-12.5 mm; upper side bronzed black; underside dark reddish brown; aedeagus with a relatively broad tip that is slightly bent downwards (Fig. 138) ... 9. *Ilybius subaeneus* Erichson (p. 57)

15. Mainly black, elongate insect; suture with a series of fine punctures on either side .... 13. *Agabus guttatus* (Paykull) (p. 53)

- Elytra pale brown, more rounded insects; suture without fine punctures on either side 16

16. Pronotum dark and elytra paler brown with shoulders even paler .... 16. *Agabus paludosus* (Fabricius) (p. 54)

- Pronotum and elytra the same reddish brown .... 10. *Agabus brunneus* (Fabricius) (p.53)

17. Each elytron with a distinctive wavy yellow mark just behind the shoulder and other yellow markings on otherwise black elytra (Fig. 139); head predominantly red .... 6. *Agabus undulatus* (Schrank) (p. 51)

- Elytra with at most pale spots on the rear third; head mainly black 18

18. The forward of the elytral spots an 'H' shape made up of two pale spots joined together (Fig. 140) ....
.... 12. *Agabus didymus* (Olivier) (p. 53)

- The forward of the elytral spots, if they are visible, singular 19

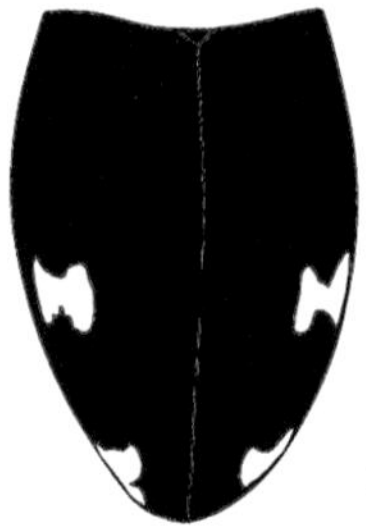

140

19. When viewed from above, the raised rim midway along side of the pronotum is thinner than the width of an antennal segment (Fig. 141) 20

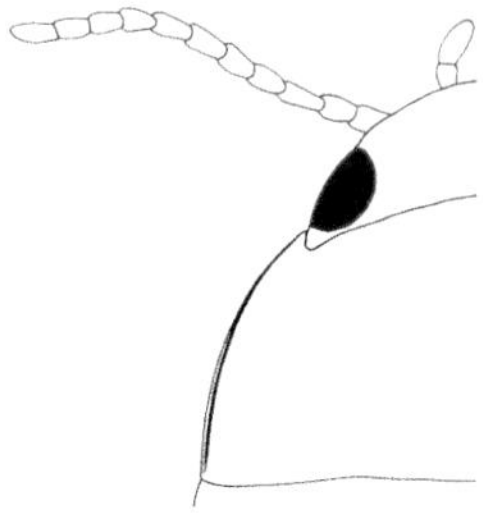

141

- When viewed from above, the raised rim midway along side of the pronotum about as wide as an antennal segment at its insertion point in the next one (Fig. 142) 23

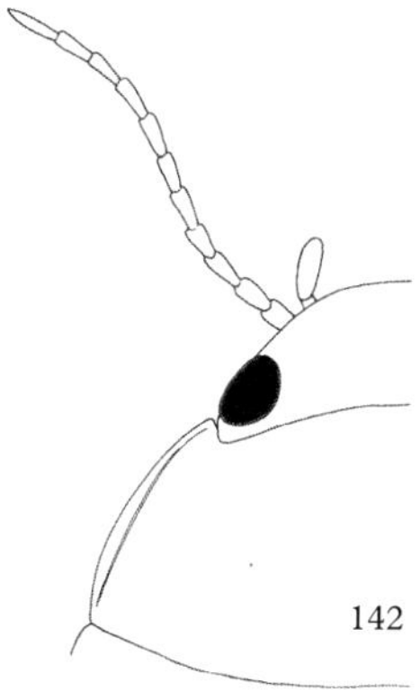

142

20. Pronotum with wide yellow sides often extending across the pronotum apart from a narrow longitudinal line (Figs 143 and 144 show the extremes); aedeagus with a hook-like tip (Fig 145) .... 1. *Agabus arcticus* (Paykull) (p. 50)

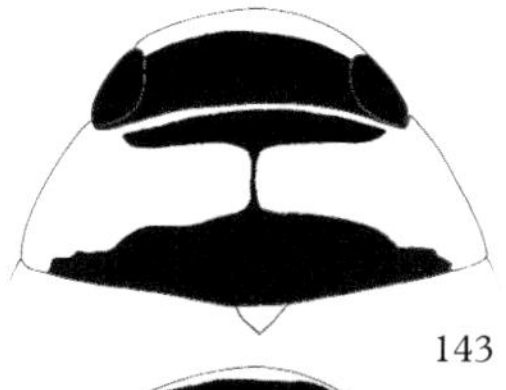

- Pronotum black; aedeagus variously shaped, sometimes hooked 21

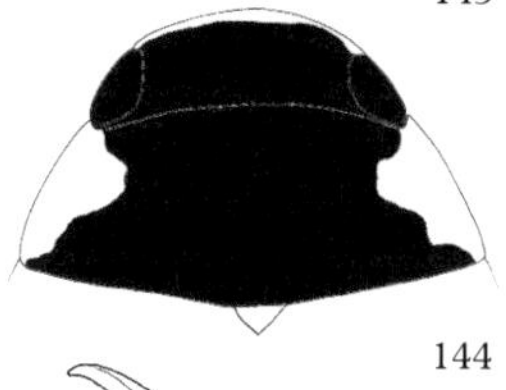

143

144

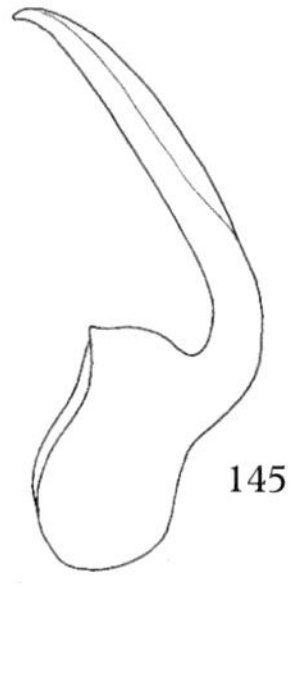

145

21. More than 8 mm long; elytra with an ill-defined crescent-shaped pale area visible if the elytra are lifted (Fig. 146); aedeagus with fine teeth visible at × 80 (Fig. 147) ....
.... 14. *Agabus melanarius* Aubé (p. 53)

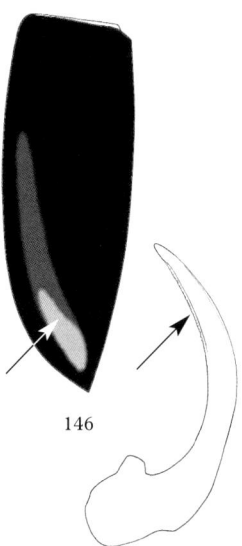

146

147

-   7.5 mm long or less; elytra each with two pale spots, one at the rear, one about halfway (Fig. 148); aedeagus without fine teeth as above 22

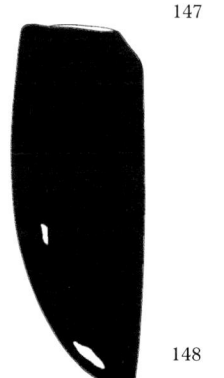

148

22. Aedeagus with a distinct tooth on the underside (Fig. 149); last antennal segment darkened in both sexes, the rest usually pale; tip of maxillary palp pale; elytra parallel-sided ....
.... 7. *Agabus affinis* (Paykull) (p. 52)

-   Aedeagus with underside smoothly curved (Fig. 150); 7th to 11th antennal segments and last segment of maxillary palp darkened; elytra slightly curved at sides to give a more rounded appearance than *affinis* .... 18. *Agabus unguicularis* (Thomson) (p. 54)

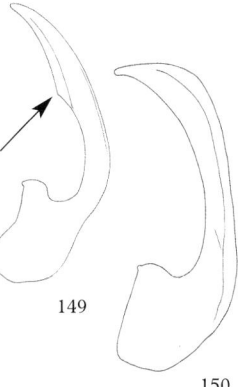

149

150

23. Metacoxal lines not quite meeting the metasternum (Fig. 151); bronze insects 6.9-8.7 mm long 24

-   Metacoxal lines joining to the metasternum; bronze or black insects 5.5-11.0 mm long 25

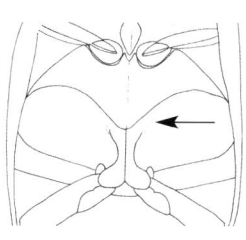

151

24. Parameres of male genitalia with strong bristle pads at tips, often visible without complete dissection (Fig. 152); antennal darkening (both sexes) usually confined to last segment; 7.5-8.7 mm .... 3. *Ilybius chalconatus* (Panzer) (p. 55)

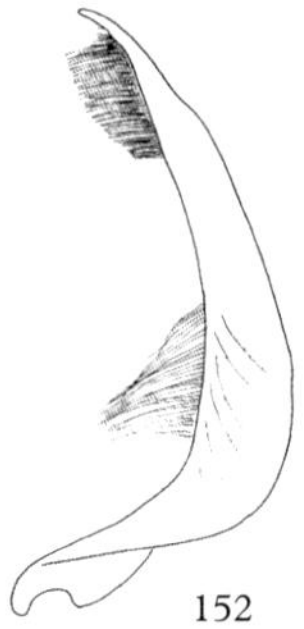

152

– Parameres with flimsy ends, sometimes visible without complete dissection (Fig. 153); antennal segments 7-11 darkened in both sexes; 6.9-8.5 mm .... 7. *Ilybius montanus* (Stephens) (p. 56)

Females with dark antennae could be either *chalconatus* or *montanus*.

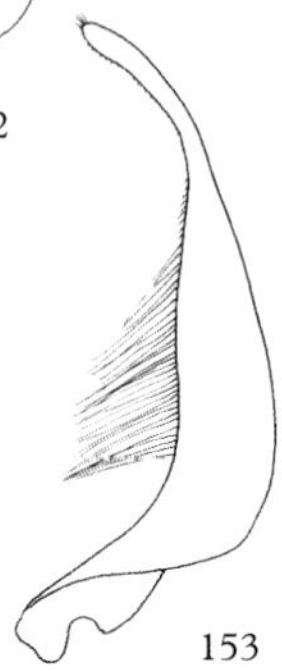

153

25. Only the last antennal segment darkened; male inner fore claw with a tooth (Fig. 154); not less than 8 mm long .... 8. *Agabus biguttatus* (Olivier) (p. 52)

154

– Antennal segments 5-11 partly darkened; male inner fore claw without a tooth, possibly with a blunt expansion (see couplet 27 Figure 156) ; not more than 8 mm long 26

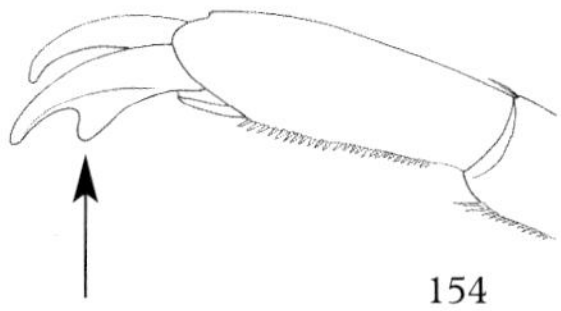

26. Pronotum almost entirely black, only the raised rim paler; aedeagus with a double point (Fig. 155); length 6.6-8.0 mm .... 2. *Agabus congener* (Thunberg) (p. 50)

– Pronotum with pale margins extending beyond the raised rim; aedeagus with a simple point 27

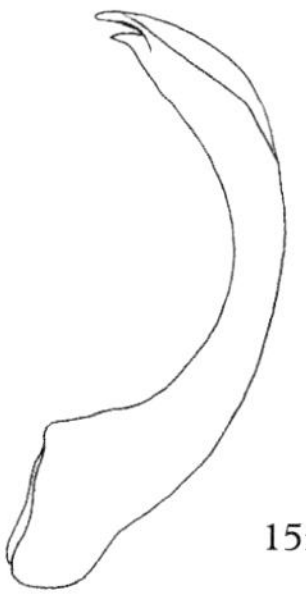

155

27. Males with inner fore claws with blunt lobes (Fig. 156); femora of fore legs not thickly haired; aedeagus with a large base (Fig. 157) and parameres correspondingly robust, with hairy tips (Fig. 158). Both sexes: body broadly oval and elytra bulging such that the edges of the epipleurs cannot be seen from above simultaneously on both sides for much of the elytral edge (Fig. 159) .... 5. *Agabus uliginosus* (Linnaeus) (p. 51)

-   Males (with thickened fore and mid tarsi with sucker hairs): fore claws without lobes; femora of fore legs with a thick line of hairs Fig. 160); aedeagus with a long narrow extremity and with the base not as enlarged as above (Fig. 161); parameres with a naked mobile appendage (Fig. 162).
    Both sexes: body elongate oval and elytra flat enough for the edges of the epipleurs to be seen when viewed directly from above on both sides for much of the elytral length (Fig. 163) .... 4. *Agabus labiatus* (Brahm) (p. 51)

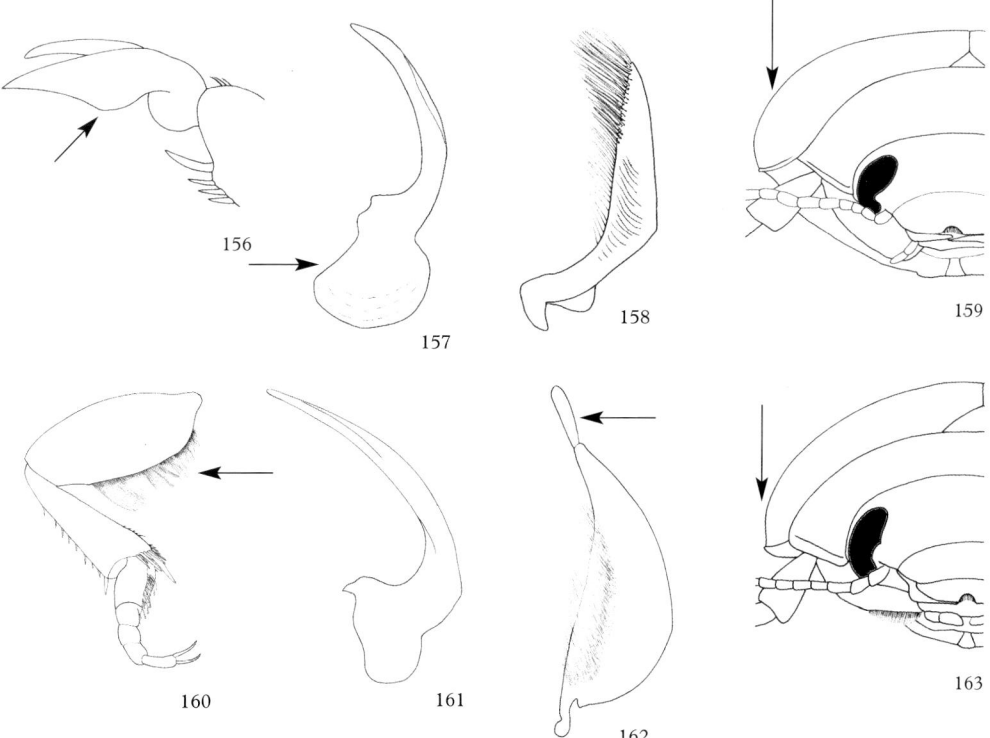

49

1. *AGABUS* Leach (see Key 7, p. 42)

The most speciose of the medium-sized genera of Dytiscidae, though reduced in size following removal of several species to *Ilybius* as redefined by Nilsson (2000). The current arrangement of subgenera might be viewed as the most recent of ongoing attempts to clarify relationships within this group. These medium-sized diving beetles include several of the most abundant water beetles in Britain, found in both stagnant and running water habitats.

Subgenus *Acatodes* C.G. Thomson

1. *Agabus arcticus* (Paykull) Plate 35

Length 6.7-8.3 mm. Most likely to be mistaken for *A. sturmii*, which can occur in the same habitats. The majority of *A. arcticus* have largely pale pronota through which the prothoracic glands can be seen as a white band in life. *A. arcticus* lives in permanent, still, montane and/or northern lakes and pools, usually in association with marginal vegetation. Widespread on the Scottish mainland and major islands, even on low ground in acid lakes in Dumfries and Galloway, in England confined to mountainous parts of the Lake District and along the Pennines to the Peak District, and in Wales known only from Snowdonia. In Ireland this species occurs on mountains in Connemara and Wicklow, and is more widespread in the north. It is not known from the Isle of Man. Recorded throughout the year except January, peaking in June and August.

2. *Agabus congener* (Thunberg) Plate 36

Length 6.6-8.0 mm. Another species similar to *A. sturmii*, but smaller and darker. Acid fens, in particular in small hard-bottomed pools. Widespread but infrequent on the Scottish mainland and on major islands north to the Orkneys. In Wales fairly frequent in the northern mountains, but scattered otherwise in fens such as the Cors Bordeilio on Anglesey and Abercamlo Bog in Radnorshire. Infrequent in Ireland, mainly in the north but in a few localities in the centre south to Kilkenny. The English distribution follows moorland south to Derbyshire and North Lincolnshire, with outlying populations in the New Forest, in Cheshire, Hertfordshire, Staffordshire, West Norfolk, and Worcestershire, plus older records for Berkshire and Somerset. It is not recorded from the Isle of Man. Recorded throughout the year peaking in June.

3. *Agabus sturmii* (Gyllenhal) Plate 37

Length 7.7-8.9 mm. The orange edge to the pronotum and elytral shoulders is easily seen in the field, and the double reticulation provides the best character under the microscope. Specimens kept in preservation fluid should be thoroughly dried out before examination. *A. sturmii* lives in vegetated, permanent waters in lakes, ponds, and slow sections of drainage ditches and rivers. Common throughout Britain and Ireland except for the Shetlands, the Scillies and the Channel Isles. Recorded throughout the year, peaking in June and August.

Subgenus *Agabus sensu stricto*

4. *Agabus labiatus* (Brahm) Plate 38

Length 5.8-6.8 mm. A small insect with weakly bronze and slightly transparent elytra. In exposed, still waters, including acid pools such as in the New Forest, alkaline temporary water such as the turloughs of the Burren, and duneslack pools in the north of the Isle of Man, probably because such habitats do not support fish that might prey on the free-swimming larvae. This beetle is rarely found in recent man-made habitats. In Scotland mainly in the north-east from Fife to Inverness, having been lost from the Central Belt and the south. One part of the recent English distribution runs from north-west Yorkshire through the Vale of York to Thorne Waste, Rauceby Warren in South Lincolnshire and then to the Brecks and the Broads. The other main part is based on heathlands of Hampshire, Surrey and other Home Counties. There are also sites scattered about in the southern Lake District, Shropshire, Herefordshire, North Somerset, Oxfordshire, and East Sussex. In Ireland the distribution is compact, being centred on the western karst of Clare and south-east Galway, running out to Limerick, Longford, Meath, Offaly, and Roscommon. Given this wide distribution elsewhere the Welsh distribution is surprisingly restricted, known from two hill lakes in Breconshire and Radnorshire. *A. labiatus* also occurs on Jersey. Recorded throughout the year, peaking in April and October/November.

5. *Agabus uliginosus* (Linnaeus) Plates 39 and 40

Length 6.4-7.6 mm. A strongly rounded and globular species with the thick pronotal rim a prominent feature; the toothed fore claw of the male will settle any concerns about identity. A species of fluctuating water bodies in grassland including merse and in pools associated with intermittent streams on limestone and chalk. *A. uliginosus* has dimorphic females, the shining type form like the male and the matt form *A. dispar* (Bold, 1849). The male-like form is known from the south of England and Wales, from the levels surrounding the Severn Estuary in Somerset and Monmouthshire to Herefordshire, Wiltshire and Oxfordshire. This population would include the only record off the British mainland, for Sully Isle in Glamorgan in the 1950s. *A. dispar* occurs from East Anglia to northern Scotland with outliers, a recent record in Shropshire and older ones on the Solway and in Hertfordshire. The beetle is occasional in fens from Wood Walton Fen to the Brecks. It is then frequent from Nottinghamshire to north-west Yorkshire, becoming more scattered through County Durham to the Campfield Kettlehole in North Northumberland and Lurgie Loch Moss in Berwickshire. The northernmost records are for the Coul Links in East Sutherland and Tain in Easter Ross. Recorded throughout the year, peaking in May and August.

6. *Agabus undulatus* (Schrank) Plate 41

Length 7.0-7.8 mm. Utterly distinctive because of the elytral pattern. In permanent, well-vegetated stagnant water in fenland, usually over peat or clay. The present day distribution is a remnant of a much wider distribution in England: it falls into four areas – in and around Askham Bog in mid-west and south-west Yorkshire; in fens of Northamptonshire, Huntingdonshire, Cambridgeshire, in particular Wicken Fen, and in South Lincolnshire; in a few of the palsa scar fens and in pits on the Stanford Training Area in the Brecks; and a single record for a stream in West Suffolk in 1994. Recorded throughout the year except January, peaking in April and August.

Subgenus *Gaurodytes* C.G. Thomson

7. *Agabus affinis* (Paykull) Plate 42

Length 6.2-7.1 mm. A small parallel-sided species most similar to *A. unguicularis*, and safely distinguished on the shape of the underside of the aedeagus. Associated with flooded moss, either *Sphagnum* or *Calliergon* species in mesotrophic or acid fens, sometimes in lake edges. In the south of England the species is scattered from the extreme south-west, to the Somerset Levels, and the heathland of Dorset, Hampshire and Surrey, reaching Burnham Beeches in Buckinghamshire, and with isolated sites on the Greensand of West Norfolk. Separately, frequent in fens and bogs north of a line from the Severn to the Humber, particularly so in central and north Wales, in much of the Lake District and in south-west Scotland, but less frequent in the Highlands. Scarce on islands being known from Skomer, the Isle of Man, the Clyde Isles, Islay, Jura, Raasay, and Fladda. In Ireland *A. affinis* is common in the west and north, but appears to be largely absent from the south except near to the coast. Recorded throughout the year, with peaks in April and June.

8. *Agabus biguttatus* (Olivier) Plate 43

Length 8.4-9.0 mm. Most likely to be confused with *A. guttatus*, but the darkened extremities of the antennae and the toothed male fore claw distinguish it safely. Found in lime-rich waters, and considered to be semisubterranean in Britain, living in the twilight zone of springs and under culverts, flushed to the surface by heavy rain when it may even be encountered on marine beaches. The distribution loosely tracks the availability of lime-rich running water across Britain, being absent from much of East Anglia and the coastal land north to County Durham, where it is frequent in association with Magnesian Limestone, as it is along the Pennines. Other strongholds are the chalk north and west of London and the marl beds of Cheshire. Absent from the Scottish Highlands and Islands, with outlying populations on metamorphic limestone in Moray and Banff, and Kintyre. There are old records for the Isle of Man. *A. biguttatus* is rare in Ireland despite the abundance of limestone, with recent records from Antrim, mainly on the chalk, East Donegal, Fermanagh, Meath, North Tipperary, Tyrone, and Waterford. Recorded throughout the year, peaking in July.

9. *Agabus bipustulatus* (Linnaeus) Plate 44

Length 9.5-11.6 mm. A large black species with which one quickly becomes familiar in the field, the females being dull and the males shiny, but both having elongate sculpture on the elytra from which water runs off in streaks. Common in almost any stagnant water habitat other than lake shores, also in trickles and backwaters of large running water bodies. This is the commonest diving beetle in Britain and Ireland, ranging to the most remote islands, and including Alderney, Guernsey, Herm, Jersey and Sark. Gaps in the distribution are some of the driest parts of England but include areas such as North Lincolnshire where there has been a systematic survey of drainage ditches. This species has a northern and montane form *solieri* Aubé, 1837, smaller than the type and with a narrow pronotum: this is considered to be no more than the extreme of a clinal variation, with many lowland Scottish sites having a mixture of both forms. Another variant is the form with parts of the elytra red, usually to be found in the montane *solieri*, but also in more normally sized beetles in lowland dystrophic pools. Recorded throughout the year, peaking in June.

10. *Agabus brunneus* (Fabricius) Plate 45

Length 7.6-9.1 mm. This brown or reddish brown species is only likely to be confused with teneral specimens of other stream-living species. Mainly in lowland streams associated with heathland and near the south coast, interstitial where the substratum is coarse gravel. Confined to West Cornwall, Dorset, South Wiltshire and South Hampshire. Recorded in January, February and April-October, with peak occurrences in September.

11. *Agabus conspersus* (Marsham) Plate 46

Length 7.0-8.3 mm. One of the most distinctive species, pale without the pronotal spots of *A. nebulosus*. Largely confined to brackish water, usually amongst sparse vegetation in coastal lagoons and ditches. More or less continuous around the coast from Dorset to south-east Yorkshire, well established in County Durham, where it has been found not only in tidal water but also a little inland in old mine workings with a low chlorinity but a high sodium content. There are doubtful records further to the north in South Northumberland, East Lothian and North Aberdeenshire. The western distribution in Britain includes Cornwall, the north Devon coast, the Severn Levels, and on the Solway, where this species can be common. There is an inland record for Nottinghamshire. The Irish distribution is, with the exception of an individual found in a moorland lough in Galway in 1921, confined to the south and east coast from Mid Cork to Louth, and with two records from 1909 for County Down. Recorded throughout the year except February, peaking in June and August.

12. *Agabus didymus* (Olivier) Plate 47

Length 7.5-8.0 mm. This bronze insect is easily recognised in the field because of its elytral spots. Typically in slow running water, in rivers, streams and drainage ditches, occasionally on exposed peat and in gravel and sand pits. Found throughout England north to a line from South Lancashire to Berwickshire, where the only Scottish record is from the Buskin Burn at Coldingham in 1939. The Welsh distribution is largely coastal and in the Marches. Island records are for the Isle of Wight and Anglesey. Recorded throughout the year, peaking in May and August.

13. *Agabus guttatus* (Paykull) Plate 48

Length 7.8-9.2 mm. In shallow running water mainly in headwaters. Common over much of Britain, Ireland and the Isle of Man where the habitat is available. Thus there are gaps, particularly around the Wash. *A. guttatus* is found on all major Scottish islands, extending to the Shetlands including Fair Isle and to St. Kilda. This species ranges across much of eastern Ireland, but is more coastal in the west, with no records from the south coast apart from one site in Waterford. Recorded throughout the year, peaking in June.

14. *Agabus melanarius* Aubé Plate 49

Length 8.5-9.8 mm. Likely to be confused with *A. bipustulatus* or *A. guttatus* in the field, but it can be distinguished by lifting an elytron to see the vague pale marking, as opposed to the two distinct spots on each elytron of the other two species. Confined to small, partly shaded pools receiving seepage, usually on the sides of wooded hills. In Scotland known from Argyll in deer wallows in native woodland and in coniferous forest from Inveraray to above Lough Creran, with a 19th Century record from Orkney. The English distribution is patchy with major centres based on afforested hill land in Yorkshire, mainly in the north-east but also recorded from the north-west and south-west, Derbyshire and Staffordshire, South Devon, and in and around the Weald in Berkshire, North Hampshire, Surrey, Sussex and West Kent. More isolated populations have been recorded from Cheshire, County Durham, Cumberland, Dorset, Herefordshire, South Lancashire, South Northumberland, South Somerset, Westmorland, and Worcestershire. Welsh records are for Caerfyrddyn, Glamorgan and Meirionydd. *A. melanarius* was originally recorded in Ireland as a single specimen in a mountain lake in Fermanagh in 2008, and has subsequently been found in numbers in hillside seepages in the vicinity. Recorded throughout the year, peaking in February/March and October.

15. *Agabus nebulosus* (Forster) Plate 50

Length 8.2-8.6 mm. The two dark spots on the otherwise pale pronotum are distinctive: specimens without these spots are very rare and may be confused with *A. conspersus*; the different patterning of the elytra will differentiate these species, those of *nebulosus* having many dark interconnected spots, those of *conspersus* being a more uniform, semi-transparent brown. In exposed stagnant water, typically in disturbed or newly created ponds. Found across much of Scotland to the Shetlands except for the Western Highlands and the inner islands, but with a western fringe of records from St. Kilda, Lewis, South Uist, Barra, Vatersay, the Monach Islands, Coll, Tiree, Islay and the extreme tip of Kintyre. *A. nebulosus* has a similar distribution in Wales with a large gap across the more mountainous parts. It is widely distributed in Ireland and England, and found in the Isle of Man, Lundy, the Scillies, Alderney, Guernsey and Jersey. Recorded throughout the year, peaking in July.

16. *Agabus paludosus* (Fabricius) Plate 51

Length 6.5-8.0 mm. A small shiny version of *A. sturmii*, but usually in a different habitat, in slow running water, often in vegetation such as beds of watercress or water crowfoot, and also in valley fens. Common across much of Britain and Ireland, ranging to the Shetlands and the Outer Hebrides. Known from Jersey. Recorded throughout the year, peaking in May and July.

17. *Agabus striolatus* (Gyllenhal) Plate 52

Length 7.2-7.9 mm. This strikingly parallel-sided species might be confused with *A. guttatus*, but its elongate reticulation distinguishes it from that species. In fen carr and springfed tussock fens. Mainly in the Broads in East Norfolk but also found at Guist Common, West Norfolk and on the Thorne Moors, South-west Yorkshire. Recorded from January to July, peaking in May, and from September to November.

18. *Agabus unguicularis* (Thomson) Plate 53

Length 6.0-6.7 mm. In rich and mesotrophic fens with mosses or fine grasses, sometimes in company with *A. affinis*, with which it is frequently confused if the male genitalia are not checked. Widespread in southern Scotland, absent from much of the Highlands but reaching East Inverness-shire. Frequent in lowlands in northern England to just south of the Pennines, linking up to the Cheshire Plain and relict fens through to Radnorshire. It also occurs in fens running from Northamptonshire to East Norfolk, in the Somerset Levels, on the Lizard, in the New Forest and on the Surrey heaths. Apart from in the Marches, *A. unguicularis* is found in Wales only in Anglesey. The only other small islands recorded are the Isle of Man and Guernsey. Common in the centre and north of Ireland, rare elsewhere, absent from the south-west and around Dublin. Old records indicate a much wider former distribution across Britain and Ireland. Recorded throughout the year, peaking in May and September.

2. *ILYBIUS* Erichson, 1832 (see Key 7, p. 42)

This is the other main genus of medium-sized diving beetles apart from *Agabus*. The original members of *Ilybius* (i.e. not *chalconatus*, *montanus* or *wasastjernae*) are typically whale-backed black or bronze species, quite pungent when freshly caught, and originally separated on the basis of the unequal hind fore claws and with the parameres carrying a mass of coarse bristles. The more widely interpreted genus is defined on characters that may prove difficult to detect in some British species, hence the key combining *Agabus* with *Ilybius* and taking advantage of obvious size and colour differences. One character that may become apparent is the ovipositor, made up of two robust plates with teeth or bristles.

1. *Ilybius aenescens* **C.G. Thomson** Plate 54

Length 8.5-9.8 mm. The metallic sheen and the pungent smell will differentiate this species from *I. guttiger*. Typically found in permanent, acid, stagnant waters, particularly in loose *Sphagnum* in dubh lochans over deep water. Some early records are considered to be unreliable, probably being based on *I. guttiger*. The modern distribution is scattered and mainly western: in Scotland to the Shetlands and the Outer Hebrides; similarly in northern England as far south as Chartley Moss in Staffordshire; the same in Wales, being commonest in Snowdonia; on heathland from Dorset to Surrey and West Sussex. There are isolated sites such as Priddy in North Somerset, Lee Moor in South Devon and Molinnis Moor in East Cornwall. Recorded from February to November, peaking in June.

2. *Ilybius ater* **(De Geer)** The Mud Dweller Plate 55

Length 12.5-14.5 mm. Unique in the British and Irish fauna by virtue of the combination of size and colour. In stagnant water over vegetated mud or peat, typically in the very edge of ponds. Common across lowland Britain and Ireland, absent from the Highlands but common in Speyside and other parts of lowland Inverness-shire, reaching Caithness. It is found on Anglesey, the Isle of Man, Rathlin Island, Arran and the adjacent Holy Island, Cumbrae and Islay. Recorded throughout the year, peaking in June.

3. *Ilybius chalconatus* **(Panzer)** Plate 56

Length 7.5-8.7 mm. Separable from the more common *I. montanus* with certainty only by examining the tips of the parameres. Females with only the last antennal segment darkened can be assigned to *I. chalconatus* rather than *I. montanus* with reasonable safety but specimens with more segments darkened could be either species. Associated with temporary pools, mainly wooded in the south of Britain but sometime exposed elsewhere. A recent invader of Scotland, ranging in the east from East Sutherland to the Borders. The English distribution is also eastern, in lowlands east of the Pennines except for a small number of localities in Cumberland and Westmorland. Local in England south from Cheshire and east from South Devon. Known in Wales solely from one site in Monmouthshire. Apart from the Isle of Wight, not known from any British island. The Irish distribution is scattered, in Armagh, Fermanagh, Offaly, Waterford and Wexford. Recorded throughout the year, peaking in May.

4. *Ilybius fenestratus* (**Fabricius**) Plate 57

Length 10.0-12.0 mm. Likely to be confused with the narrower *I. fuliginosus*, which has a more distinctive yellow stripe along the sides, and *I. subaeneus*, which has a darker underside: if in doubt the narrow metasternal wing sets this *Ilybius* apart from the others. In still, permanent waters in lakes, large ponds and canals, usually associated with sparse vegetation; often to be found in angling lakes, possibly because of the protection conferred by its pungent smell. In Scotland this species is confined to lochs in Kirkcudbrightshire. The English distribution is also compact, running south of a line from Cheshire to Mid-west Yorkshire and absent from the south-west except for the Levels in North Somerset. Not reported from any islands. Recorded throughout the year, except January, peaking in August.

5. *Ilybius fuliginosus* (**Fabricius**) Plate 58

Length 10.0-11.5 mm. The narrow body form combined with the strong yellow streaks will differentiate this species from other medium-sized diving beetles. In permanent water, in vegetated rivers, drainage ditches, ponds and lakes, including brackish water. Common across most of Britain and Ireland, on most islands and into most mountain systems. Recorded throughout the year, peaking in June.

6. *Ilybius guttiger* (**Gyllenhal**) Plate 59

Length 8.7-10.0 mm. Measurement of the length of this dull black species is crucial to separate it from the larger *I. quadriguttatus*. The tip of the aedeagus is diagnostic. Found in flooded vegetation, particularly mosses, in fens, often in dense reedbeds. In Scotland confined to the south-west from Dumfriesshire to Renfrewshire, also recently on Tiree. The English distribution is localised, with large areas of occupancy in Cumberland and Westmorland, from South Northumberland to South-west Yorkshire, from South Lancashire, Cheshire, Shropshire and Staffordshire, in fens from Huntingdonshire to the Broads, from Dorset to South Essex, and within the Weald. There is a wholly separate population in Cornwall. *I. guttiger* is common on Anglesey, and is otherwise scarce in Wales, known from Caernarfon, Denbighshire, Flintshire, Pembrokeshire, and Monmouthshire. The Irish distribution is mainly in a broad band running to the west of a line from Kerry to Down. Apart from Tiree, the only small island occupied is the Isle of Man. Recorded throughout the year, peaking in June.

7. *Ilybius montanus* (**Stephens**) Plate 60

Length 6.9-8.5 mm. This species is far more common than *I. chalconatus*, from which it cannot be distinguished safely by external characters. Found in shallow water amongst flooded grasses, but sometimes also in pools with exposed peat. Common across most of Britain and Ireland, including as a feature peculiar to this species many of the very small islands – Sherkin, Lundy, Skomer, Holy Island (Arran), Little Cumbrae, Gometra, Eigg, Muck, Canna, Sanday, South Rona, Fladda, South Ronaldsay, Hoy, Flotta, Fair Isle, Yell, Fair Isle, Unst, and Foula. *I. montanus* is also known from the Isle of Man and Jersey. Recorded throughout the year, peaking in May and September.

8. *Ilybius quadriguttatus* (Lacordaire) Plate 61

Length 10.5-12.2 mm. A dull black species intermediate in size between *I. guttiger* and *I. ater*. In permanent, densely vegetated ponds, ditches and canals. The northernmost records are from Ireland, on Rathlin Island, the only island from which it has been recorded except for the Isle of Man, Anglesey and Wight. In England it reaches north to the Cumberland coast and South Northumberland. There are gaps in its predominantly low ground distribution, for example the south-west beyond the Exe and much of the chalk. The Welsh distribution is coastal, the Montgomery Canal providing an exception. The Irish distribution also has large gaps in the south-west and the north-west. Recorded throughout the year, peaking in June.

9. *Ilybius subaeneus* Erichson Plate 62

Length 10.0-12.5 mm. Most likely to be confused with *I. fenestratus*, but less metallic and darker with wide metasternal wings. Found in permanent water amongst vegetation, most often in man-made sites such as in mineral workings and in areas of mining subsidence, abandoned ornamental ponds, borrowpits and decoy ponds, but also in natural coastal pools. In Scotland found in Wigtownshire and Kirkcudbrightshire with old records for Dumfriesshire and Angus. In England widespread over the north-east and Yorkshire coalfields, extending south into Cheshire and Leicestershire. The rest of the records are scattered with modern ones to East Sussex and South Hampshire. The sole Irish record is from the Glastry claypits in County Down, and *I. subaeneus* also occurs in the Isle of Man. Recorded throughout the year except March and December, peaking in June.

10. *Ilybius wasastjernae* (C.R. Sahlberg) Plate 63

Length 6.4-7.3 mm. In the area where *I. wasastjernae* lives in Scotland this parallel-sided species is likely to be confused only with *Agabus affinis* and *A. guttatus*, from which it can most readily be distinguished by its double reticulation. The habitat is shaded pools formed beneath root-plates of trees in wet forest and in holes created when trees are windthrown. Confined to the north-facing coniferous forest of Abernethy Forest in Moray and East Inverness-shire. Recorded in May, July and October, commonest in May.

3. *PLATAMBUS* Thomson

1. *Platambus maculatus* (Linnaeus) Plate 64

Length 7.5-8.5 mm. An unmistakable species, though the darker specimens can cause confusion until the pale elytral margins are detected. Permanent rivers and streams, often found amongst submerged parts of overhanging vegetation, also on lake shores subject to much wave action. Common over much of mainland Britain, but absent from the extreme south-west, the land surrounding the Wash and other fenland lacking fast running water; absent from the Isle of Wight, Anglesey and the Isle of Man, but reaching Islay and Skye, possibly also Taransay. *P. maculatus* was discovered in Ireland in the Urrin River in Wexford in 2008. Dark specimens, probably referable to var. *inornatus* Schilsky, are confined to base-poor waters in rivers and lochs in northern England and Scotland, but also in the New Forest. Recorded throughout the year, peaking in August.

Subfamily COLYMBETINAE Erichson

4. *COLYMBETES* Clairville

1. *Colymbetes fuscus* (Linnaeus) Plate 65

Length 15.0-17.0 mm. A distinctive large species, easily recognised by its colouring and the transverse sculpture of the elytra. Common throughout lowland Britain and Ireland ranging to the Outer Hebrides and the Shetlands, also on the Scillies, Alderney, Guernsey and Jersey. Its avoidance of the uplands results in its being largely coastal in Wales and northern England and northern Scotland but more central in Ireland. Recorded throughout the year, peaking in July and September.

5. *RHANTUS* Dejean, 1833

Six species ranging in size from 9 to 12 mm long, associated with stagnant waters. The entirely black *Rhantus grapii* is assigned to a separate subgenus, *Nartus* Zaitzev, which has occasionally been given generic status. There is potential for confusion in the several name changes affecting other species, *R. suturalis* (MacLeay) being previously known as *R. pulverosus* (Stephens), whilst *R. suturellus* (Harris) was known as *R. bistriatus* auct. in Britain. A species that became extinct in Britain in the 1900s was known as both *R. adspersus* sensu Fabricius and *R. aberratus* Gemminger & Harold, but is now considered to be *R. bistriatus* (Bergsträsser).

Males are distinguished from females by the presence of pale sucker hairs on the undersides of the expanded 1st-3rd segments of the front and mid tarsi. Also, the male front and mid claws are often strongly developed. There should be no need to dissect the male genitalia.

Key 8. The species of *Rhantus*

1. Elytra and the pronotum black ....
.... 1. *Rhantus grapii* (Gyllenhal) (p. 60)

- Elytra mottled yellow and black 2

2. Pronotum without marks in the centre, dark marks if any confined to the front and hind edges though there may be faint egg-shaped marks based on muscle attachment sites seen through the transparent cuticle (Fig. 164 – *Rhantus exsoletus*); male front claws about the same length 3

164

- Pronotum darkly marked in the centre, sometimes with faint patches to the side of the main patch (Fig. 165 *R. frontalis*; or without patches – Fig. 166 *R. suturalis*); male front claws unequal in length, one no more than ⅔rds the length of the other 5

165

166

3. Underside yellow-brown or brown; male with long front tarsal claws, the longest appreciably longer than the last tarsal segment (Fig. 167) .... 3. *Rhantus exsoletus* (Forster) (p. 60)

- Underside predominantly black; male front tarsal claws as long as, or only a little longer than, the last tarsal segment 4

167

4. Abdomen all black; pronotum with distinct dark bands on front and hind margins (Fig. 168); male front tarsal claws both longer than the last tarsal segment (Fig. 169) .... 6. *Rhantus suturellus* (Harris) (p. 61)

168

169

- Abdominal segments black with yellow sides; pronotum with at most vague markings (Fig. 170); shortest male front tarsal claw similar in length to the last tarsal segment (Fig. 171) .... 2. *Rhantus bistriatus* (Bergsträsser) (p. 60)

This species is extinct in Britain!

170

171

5. 2-3 narrow yellow lines "picking their way" between the black spots on each elytron (Fig. 172); underside of male abdomen with dark segments which are paler at the sides, female abdomen *vice versa*; male with longest front tarsal claw longer than the last tarsal segment (Fig. 173); antennal segments dark-tipped; female with elytral sculpture of short vertical cuts best seen in highlights .... 4. *Rhantus frontalis* (Marsham (p. 60)

172

173

- Lines down elytra rather variable, petering out in parts, often absent or even made of coalescing dark spots; abdomen of both sexes all black (except in immature specimens); male with front tarsal claws shorter than last tarsal segment (fig. 174); antennae yellow; female elytra smooth .... 5. *Rhantus suturalis* (Macleay) (p. 61)

174

1. Subgenus *Nartus* Zaitzev

1. *Rhantus grapii* (Gyllenhal) Plate 66

Length 10.0-11.0 mm. In shallow water amongst vegetation, dead leaves or mosses in stagnant water, often partly shaded, and usually in lowlands. The largely black body can cause this species to be confused with *Agabus bipustulatus* in the field, but it is more adept than that species at hiding among debris when caught. In England the distribution follows old fenland and commonland in Somerset, Dorset, the New Forest, East Anglia and south of London but not on the Thames Marshes, with scattered records from East Cornwall, North Devon, Berkshire, Shropshire, and Derbyshire as an outlier to a large area of occupancy around the Humber. *R. grapii* occurs along the southern Welsh coast, being common in Pembrokeshire and it abounds on Anglesey. It is recorded from Jersey and the Isle of Man. The Irish distribution runs from the interdrumlin mires of Armagh and Down to Kerry with few records for the coast. *R. grapii* has been recorded throughout the year, peaking in May and August.

2. Subgenus *Rhantus* Dejean

2. *Rhantus bistriatus* (Bergsträsser) Plate 67

Length 9.0-9.5 mm. The habitat abroad is exposed temporary grassy ponds. Known only from West Suffolk, East Norfolk, Cambridgeshire, Huntingdonshire, and North Essex in the 18th Century and early in the 19th Century. The last specimen was taken in the Norfolk Broads in September 1904. If rediscovered its relatively small size and the pale pronotum associated with a dark underside should attract attention.

3. *Rhantus exsoletus* (Forster) Plate 68

Length 9.0-10.0 mm. *R. exsoletus* is easily recognised in the field by its entirely yellow underside and largely yellow pronotum. It is found along lowland lake, reservoir and pond edges amongst vegetation, usually in acid or mesotrophic conditions. Frequent in large tracts of fenland and heathland but largely absent from coastal levels. In England it is absent from much of the south-west and the lower Weald. The Welsh distribution is largely northern: on Anglesey it is known only away from the calcareous fens and levels. It is common in southern Scotland but rare in the Highlands, and has been recorded from Arran, Bute, the Cumbraes, Fladday, Rum, Scalpay, Skye, South Rona, and South Uist. It is recorded from Jersey but not from the Isle of Man nor from the Isle of Wight. In contrast the Irish distribution includes much of the karst, and Tory and Rathlin Islands. *R. exsoletus* has been recorded throughout the year, peaking in June and August.

4. *Rhantus frontalis* (Marsham) Plate 69

Length 9.4-11.4 mm. The strong tripartite dark markings on the pronotum should set this species apart, and the pale markings on the abdomen also provide a good character in the field. In lowland pools amongst vegetation, often over partly exposed substrata, in particular sand. *R. frontalis* is restricted in its distribution in England to East Anglia, the Thames Marshes, the Romney Levels, the Dorset Heaths and the Somerset Levels. In Wales it is found only on Anglesey. In Scotland this species has recently expanded its distribution and is currently known from Wigtownshire, Kirkcudbrightshire, Lanarkshire and West Lothian, with older records north-west to Ayrshire and north-east to Angus. It is known from the north of the Isle of Man and ranges over central and coastal Ireland. *R. frontalis* has been recorded throughout the year except December, peaking in May and August.

5. *Rhantus suturalis* (Macleay) The Supertramp Plate 70

Length 10.0-13.0 mm. This is the largest British and Irish *Rhantus*, the smudged central mark on the pronotum being an important field character. The common name stems from Balke *et al.* (2009), who demonstrated that this species originated in New Guinea 1.5 million years ago, becoming the most cosmopolitan water beetle ranging from New Zealand to Ireland and Caithness. Usually in lowland stagnant waters, often recently created and polluted. This species occurs across the whole of England including the Isle of Wight and Lundy. In Wales it is still largely confined to the south. In Scotland it is still expanding its range, having reached Caithness in the east and Glasgow in the west. In Ireland it was found in Belfast in the 1930s but has recolonised Ireland from the south-east, having recently reached Roscommon and Down. Similarly it has been found in the Isle of Man in 2009, on Alderney in 2011, and has been known on Jersey since 1992. *R. suturalis* has been recorded throughout the year, peaking in May and September.

6. *Rhantus suturellus* (Harris) Plate 71

Length 9.4-11.0 mm. Typical of upland peat pools and lochans, particularly associated with those thinly vegetated with cottongrasses. *R. suturellus* is frequent in northern England and on heathland from Dorset to Berkshire, with isolated recent records from Northamptonshire and West Norfolk. In Wales the distribution is mainly montane, but includes Anglesey. *R. suturellus* is also frequent in Scotland, known from the principal islands in the Hebrides the Orkneys and Shetlands, including Foula and Unst. It is also found on the Isle of Man and in mainly montane parts of Ireland. *R. suturellus* has been recorded throughout the year except December, peaking in July.

Subfamily COPELATINAE Branden

6. *LIOPTERUS* Dejean

1. *Liopterus haemorrhoidalis* (Fabricius) The Piles Beetle Plate 72

Length 6.3-7.9 mm. Easily recognised as *Agabus*-sized, but narrow and red. In richly vegetated lowland ponds and ditches, usually with mosses, often in cool waters, either shaded or spring-fed. Common in England north to Yorkshire in the east and Westmorland in the west, largely in the Marches and coastal in Wales, on the north of the Isle of Man, in Scotland along the coast from the Mull of Galloway to Gatehouse of Fleet, on the south coast and in the west of Ireland north to Galway, with only one known site in the north in Lackan Bog, County Down. Also known from Alderney, Guernsey and Jersey. Recorded throughout the year, peaking in April and September.

Subfamily DYTISCINAE Leach

7. *ACILIUS* Leach

Two large species associated with stagnant waters. Broadly oval and widest behind the middle, quite flattened. Pronotum pale with two transverse bands connected together at their extremities. Underside usually black in parts.

Males are distinguished from females by the widened 1st-3rd segments of the front tarsi forming a circular pad provided with sucker hairs typical of the Dytiscinae. The females have each elytron with four hair-filled grooves. There should be no need to dissect the male genitalia.

Key 9. The species of *Acilius*

1. Hind femora pale (Fig. 175); abdominal sternites with more pale than darkened areas, sometimes whole underside pale; weak depressions on pronotum of female free from hair; less than 16 mm long .... 1. *Acilius canaliculatus* (Nicolai)

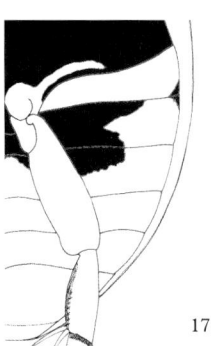

175

- Hind femora with dark marking on at least inner halves (Fig. 176); all but last of the abdominal sternites predominantly dark, pale markings confined to spots near their outer edges and along the rear edges; pronotal depressions of females with long hairs; usually 16 mm or more .... 2. *Acilius sulcatus* (Linnaeus)

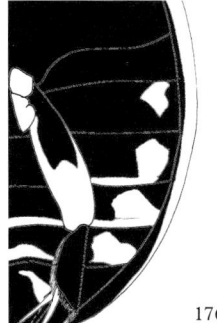

176

1. *Acilius canaliculatus* (Nicolai) Plates 73 and 74

Length 14.0-15.5 mm. This species is typical of peat cuttings, not necessarily steep-sided as for *A. sulcatus*, but often in shallow, well vegetated pools and fens, sometimes in part shade. Patchily distributed in England, from the Kent and Sussex Levels and associated woodland, Berkshire, the Cheshire Plain, Thorne Wastes and surrounding lowlands in Yorkshire and North Lincolnshire. It is also known from Shropshire in two sites, one, Whixall Moss running into the only known site in Wales, Fenn's Moss in Denbighshire. *A. canaliculatus* is scattered around the Lake District and more frequent in the Scottish Borders, also in Ayrshire and Glasgow, and then in eastern Scotland north to Speyside and the Black Isle, but not known from any true Scottish island. This species is frequent in the west, centre and the north of Ireland, with outlying records from Waterford, but it is not known from the south-west. Not known from the Channel Isles. Recorded throughout the year except January, peaking in July; larvae have been recorded from May to August.

2. *Acilius sulcatus* (Linnaeus) Plate 75

Length 15.7-18.0 mm This species is most typical of steep-sided pools, often ranging into deep and clear water in the absence of fish. It is known throughout Britain and Ireland, north to Shetland. The English part of its distribution is mainly lowland whereas it is more typical of isolated fishless upland lochans in Scotland, where it is usually darker (var. *scoticus* Curtis, 1828) than the lowland form. Scattered across Ireland, less frequent than *A. canaliculatus*. It is known from most islands, including the Isle of Man, Rathlin, Lundy, the Isle of Wight, Guernsey, Herm, and Jersey. Recorded throughout the year, peaking in June and September; larvae have been recorded from April to August, peaking in June.

8. *GRAPHODERUS* Dejean

Two extant and one extinct species associated with stagnant waters. Oval or pear-shaped species 12-16 mm long. Pronotum with dark bands along the front and hind edges separated by a wide orange area. Elytra with a black mesh surrounding small orange spots ("spangles"). Underside orange-brown, never partly black.

Males are distinguished from females by the widened 1st-3rd segments of the front tarsi forming a circular pad provided with sucker hairs typical of the Dytiscinae, the females otherwise being similar in appearance to the males without modified elytra. There should be no need to dissect the male genitalia but it is advisable to examine the sucker hairs on the mid tarsi.

Key 10. The species of *Graphoderus*

1. Pronotum with narrow pale bands in front of the foremost of the black transverse stripes and behind the rearmost one (Fig. 177), at least at its sides; mid-tarsus of male with sucker hairs in irregular rows (Fig. 178) ....
.... 3. *Graphoderus zonatus* (Hoppe) (p. 64)

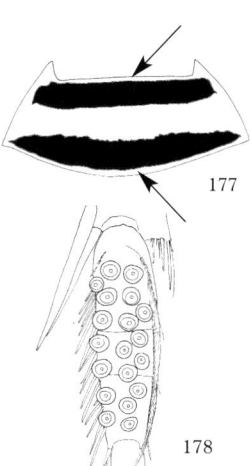

- Pronotum lacking a pale band on the hind edge of the pronotum (Figs 180 and 181) and sometimes without one on the front edge; male mid-tarsus with two regular rows of sucker hairs (Fig. 179) 2

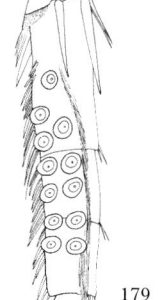

2. Intermediate pale stripe on pronotum about the same thickness as the rearmost dark stripe (Fig. 180); body only slightly wider behind the middle (see Plate 77) and not especially flattened, without sharp edges .... 2. *Graphoderus cinereus* (Linnaeus)

180

- Intermediate pale stripe on pronotum thicker than both dark stripes combined (Fig. 181); body widest well behind the middle, pear-shaped (see Plate 76), flattened, with distinctly flared and sharp-edged elytra .... 1. *Graphoderus bilineatus* (De Geer)

181

Last recorded in 1910 in England.

1. *Graphoderus bilineatus* (De Geer) The Chequered History Beetle Plate 76

Length 14.0-15.7 mm. The flared body shape, more like an *Acilius* than the other *Graphoderus*, should alert one to check on the possibility that this species has been found. Known with certainty from East Norfolk from 1906 and 1910 in the vicinity of Catfield Fen, specifically in the landward end of a drain running into the River Ant. Individuals were recorded in April and September. This species is unique amongst the water beetles of Britain and Ireland in being listed in Annexes II and IV of the Habitat Directive 92/43 EEC following its listing under the Bern Convention.

2. *Graphoderus cinereus* (Linnaeus) Plate 77

Length 13.8-15.3 mm. The width of the main pale band on the pronotum may not be enough to separate this species from *G. zonatus*, the suckers of the mid tarsus of the male providing the best character. Associated with well-vegetated stagnant waters in lowland pools and fenland drainage dykes. This species has only been recorded in England, with modern records from Dorset, Middlesex, East Sussex, East Kent, Surrey, North Essex, East Suffolk and East Norfolk, and older records for South Essex and Hereford. Recorded from April to November, with peaks in May and September.

3. *Graphoderus zonatus* (Hoppe) The Spangled Diving Beetle Plate 78

Length 12.0-15.6 mm. This species is known only from England in North Hampshire in Woolmer Forest, in the main lake and in subsidiary man-made pools. It was first found there in the area in 1953 and, judging from occasional abundance of its larvae, benefited from later excavation of the lake. Recorded in January and from April to September, with peaks in May and August; larvae have been recorded in May and from July to September.

9. *CYBISTER* Curtis

The infrequency with which these large beetles are found in England has already resulted in their being mistaken for *Dytiscus*. The pear-like body shape and the very short hind tibiae with their massive spurs should be sufficient to avoid this mistake.

Males have widened front tarsi as in other Dytiscinae. The females have smooth elytra as in the males.

1. *Cybister lateralimarginalis* (De Geer) Plate 79

Length 29-37 mm. A species typical of well vegetated, small lakes, also to be found in ditches and ponds. It was originally found in three localities in Essex from 1826 to 1836. A female was found dead beside a ditch in Leighton Moss, West Lancashire in September 2005. The 19th Century specimens were found in August and September.

10. *DYTISCUS* Linnaeus

The size of the Great Diving Beetles makes them instantly recognisable as a genus but there can be problems in identifying individual species. Their colouring varies considerably and it is advisable to use as many characters as possible in coming to a decision as to a specimen's identity – the shape of the metacoxal processes, easily seen in the field, provides the most definitive character.

Males are easily distinguished from females by the development of the fore tarsus into large circular pads. Females of four of the species always have grooved elytra but females of *D. circumcinctus* and *D. circumflexus* can have either grooved or smooth elytra. There is no need to dissect specimens in order to identify *Dytiscus* species.

Key 11. The species of *Dytiscus*

1. Underside black or generally brown; pronotum with lateral margins broadly yellow, front and rear margins only narrowly if at all yellow; tips of metacoxal processes rounded (Fig. 182) 2

- Underside predominantly yellow, sometimes with black markings; pronotum with front and side margins broadly yellow; usually also the rear margin; metacoxal processes variously shaped, but sharply pointed if the rear yellow margin is narrow 3

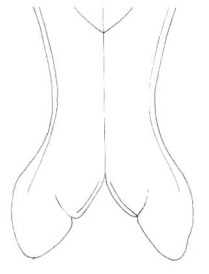

182

2. Whole underside black including the mesosternum and metasternum; less than 32 mm long ....
.... 6. *Dytiscus semisulcatus* Müller (p. 67)

- Underside typically reddish brown; 32 mm or more ....
.... 3. *Dytiscus dimidiatus* Bergsträsser (p. 67)

3. Metacoxal processes bluntly pointed 4

- Metacoxal processes sharply pointed (Fig. 183) 5

4. Hind margin of pronotum with a yellow band about as wide as the lateral bands .... 5. *Dytiscus marginalis* Linnaeus (p. 67)

- Hind margin of pronotum without a yellow band ....
.... 3. *Dytiscus dimidiatus* Bergsträsser (p. 67)

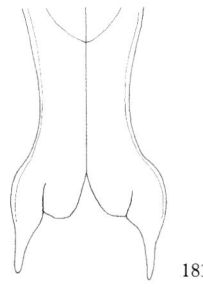

183

5. Not exceeding 30 mm in length; pronotum with very broad yellow margins around its entirety; elytral colouring typically made up of many fine coalescing dark lines fading towards the yellow margins to less organised lines of spots (Fig. 184); scutellum with at least its centre pale; females always with grooved elytra .... 4. *Dytiscus lapponicus* Gyllenhal (p. 67)

184

- Typically 30 mm or more in length, rarely down to 27 mm long; pronotum with narrow yellow bands, lateral margins wider than front and hind edges; elytra uniformly dark except for lateral yellow margins; scutellum dark or yellow; some females smooth like the males 6

6. Underside mainly yellow but most plates with edges widely black, also much of the metasternum, metacoxal plates and last visible sternite dark; eyes without a pale rim; scutellum with at least its middle yellow .... 2. *Dytiscus circumflexus* Fabricius (below)

- Underside pale yellow with only narrowly darkened edges; eyes with a pale rim (Fig. 185); scutellum dark .... 1. *Dytiscus circumcinctus* Ahrens (below)

185

1. *Dytiscus circumcinctus* **Ahrens** Plates 80-82

Length 27.0-32.0 mm. Best distinguished by the pale underside combined with the sharply pointed metacoxal processes. The habitat is vegetated, permanent, still water in lowland ponds, lakes and drains. In England most frequent in the Midlands, ranging north to North-east Yorkshire and South Lancashire and south to Worcestershire, otherwise scattered on coastal levels in Somerset, Sussex, Kent, Suffolk and Norfolk, having disappeared from north of London and the main area of old fenland in Cambridgeshire and Huntingdonshire. The Welsh distribution is confined to Monmouthshire and Denbighshire. Frequent in Ireland mainly in the non-coastal lowlands south to Limerick and South Tipperary, north to Down. Both smooth and grooved forms of female can occur together. Recorded from February to October, peaking in May and July.

2. *Dytiscus circumflexus* **Fabricius** Plate 83

Length 26.0-32.0 mm. The crisp black markings on the underside, sharply contrasting with the yellow ground colour (inspiring one common name, The Wasp), combined with the sharp metacoxal processes, should set this species apart. The recorded distribution of this species began to change in 1982, before which it was confined to south of a line from the Humber to the Severn, apart from a 19th Century record for Glamorgan. By 2003 it had reached Anglesey, the Isle of Man, Scotland in Kirkcudbrightshire, and was well established across the English Midlands. It was first noted in Ireland in 1983, from Wexford and Roscommon, and has subsequently been found in Waterford, Limerick, Clare, South-east Galway, and Offaly. This change in distribution appeared to be associated with a change in habitat preference; the old distribution was largely confined to brackish coastal pools and ditches but the new sites colonised were lowland pools with mineralised, even polluted, water often well away from the coast. Both forms of female occur together over parts of the former range but the expanded distribution is associated with the grooved form of female. Recorded throughout the year, peaking in May and July.

3. *Dytiscus dimidiatus* Bergsträsser Plates 84 and 85

Length 32.0-38.0 mm. This is the largest British species of *Dytiscus*, usually recognised by the rich brown colour of its underside, and by only the sides of the pronotum having yellow margins. It shares with *D. semisulcatus* and *D. marginalis* blunt metacoxal processes. The distribution is centred on ancient fenland along the Severn in Somerset and Monmouthshire, on the Romney Marsh and its hinterland in Kent and Sussex, on the fens of Cambridgeshire and Huntingdonshire, and on the Broads in East Norfolk. *D. dimidiatus* appears to have been lost from fenland in Yorkshire. Other records are scattered, some such as those for the well surveyed Thompson Common in 2001 indicating recent mobility. The northernmost site was in a saltmarsh in County Durham in 1969 and the westernmost in old china clay workings in East Cornwall in 1972. However, the typical habitat is in permanent, richly vegetated drains. Recorded in every month except December, peaking in May and August.

4. *Dytiscus lapponicus* Gyllenhal Plate 86

Length 22.0-28.0 mm. The typical habitat is in the fishless dubh lochans of tension pool complexes in the Flow Country or in remote mountain lochs either without fish or with sufficient vegetation to provide a refuge from fish predation. In Scotland this species occurs across the Highlands south to Kintyre but is not found in the Southern Uplands. Islands from which it has been recorded are: Arran, Eigg, Harris and Lewis, Hoy, Islay, Jura, Mull, Orkney Mainland, Raasay, Rum, the Shetland Mainland, Skye, Soay, the Uists, and Ulva. In Wales it is known with certainty only from a single lake, an upland reservoir in Caernarfon. The Irish distribution is coastal, following mountain systems in South and North Kerry, West Cork, Clare, South-east and West Galway, Sligo, West Donegal, Fermanagh, Tyrone and Antrim. Recorded from April to October, peaking in June and September.

5. *Dytiscus marginalis* Linnaeus The Great Diving Beetle Plates 87 and 88

Length 26.0-32.0 mm. The typical habitat is small ponds but it is so common that it can occur in almost any aquatic habitat, and is also regularly found at light and on reflective surfaces. This species is about twice as common as the next, *D. semisulcatus*. It ascends to over 500 metres above sea level in the Cairngorms, well above altitudes at which it can be found associated with *D. lapponicus*. Although it ranges across the whole of Britain and Ireland, and occurs in Guernsey and Jersey, *D. marginalis* is scarce in the northern isles, and is not recorded from the Orkneys and Shetlands. Recorded throughout the year, peaking in May and September.

6. *Dytiscus semisulcatus* Müller The Black-bellied Diving Beetle Plates 89 and 90

Length 22.0-30.0 mm. This species tends to occupy shallower stagnant water habitats than *D. marginalis*, and it also frequents sluggish streams. It ranges across the whole of Britain and Ireland to the Shetlands and the Outer Hebrides, and is also found on Jersey. This species is common in East Anglia but is typically more western than *D. marginalis*. It is common across Ireland. Recorded throughout the year, peaking in April and September: the distinctive larva has been recorded in all months except August, October and November with a strong peak in April.

11. *HYDATICUS* Leach

The two British species are easily recognised as small black dytiscines, the males having the front tarsi expanded as in *Dytiscus*. There should be no need to dissect the males.

Key 12. The species of *Hydaticus*

1. Elytra without pale bars on the shoulders; black crescent extending three-quarters of the way across the pronotum ....
   .... 1. *Hydaticus seminiger* (De Geer)

- Elytra with transverse pale bars on the shoulders; pronotum with a black crescent-shaped mark extending forwards about half way
   .... 2. *Hydaticus transversalis* (Pontoppidan)

1. *Hydaticus seminiger* (De Geer) Plate 91

Length 13.0-14.5 mm. Once it is realised that the beetle is a small dytiscine, with the cup-shaped front tarsi, rather than a large agabine, then there is no difficulty in recognising this species: apart from the margins the elytra are wholly black, though very occasionally a few faint yellowish marks can be seen, but never the strong transverse line seen in *transversalis*. Associated with permanent standing waters, usually amongst dense vegetation or debris in partly shaded sites. The abandoned peat cuttings of Ireland and the woodland floating bogs of the Weald are particularly favoured. It is often found on the coast but avoids brackish water. In England it is frequent south-east of a line from Dorset to Huntingdonshire and Norfolk, otherwise frequent on the Somerset Levels and on the Cheshire Plain extending into Wales in Denbighshire and Flint, with outliers in Shropshire, Staffordshire and Worcestershire. There is one 19th Century record for Yorkshire at Askham Bog. The Irish distribution is remarkable, the earliest record being for Killarney in 1929 after which it was not found again in North Kerry until 2007, with as many records as for England in the intervening period, reaching the Galway island of Inishbofin in the west, Waterford in the south-east and Belfast Bay to the north; thus this species occurs 140 km further north in Ireland than on the British mainland. Known from Jersey. Recorded in all months except December, peaking in May and September.

2. *Hydaticus transversalis* (Pontoppidan) Plate 92

Length 12.0-13.0 mm. A distinctively striped species associated with rich fen in lowland ponds and fenland drainage ditches, typically in more exposed sites than *H. seminiger*. Common across the Severn marshes in the Somerset and Monmouthshire Levels and in the Broads in East Norfolk. Recently rediscovered in the Cambridgeshire Fens, also known recently in fenland in Northamptonshire, Huntingdonshire and South Lincolnshire. It was last found in Askham Bog, the only Yorkshire site, in 1930, and in North Lincolnshire in 1938, and there are two records from the 1960s for the Scilly Isles. Recorded from February to November, peaking in May and August.

Subfamily HYDROPORINAE Aubé

12. *BIDESSUS* Sharp

Key 13. The species of *Bidessus*

1. Elytra yellow with brown transverse bars, long and parallel-sided; the grooves on the elytra that match those on the pronotum extend about halfway, certainly into the second dark band, sometimes beyond it (Plate 93) .... 1. *Bidessus minutissimus* (Germar)

-   Elytra brown, short and body rounded in outline; the elytral grooves matching those on the pronotum reaching less than a quarter the length of the elytra (Plate 94) ....
.... 2. *Bidessus unistriatus* (Goeze)

1. *Bidessus minutissimus* (Germar) Plate 93

Length 1.5-1.9 mm. Associated with permanent water, usually free of any vegetation except roots, in sand or fine gravel at the edges of rivers and their side pools, and quarry pools. Last recorded in Scotland in 1991 in the River Nith; known previously from the Water of Luce, the River Ken, and the River Annan. *B. minutissimus* also can no longer be found in the Sulby River on the Isle of Man. In Wales the species has been recently recorded from the Tywi in Caerfyrddyn, the Rheidol and Ystwyth in Ceredigion, the Dyfi in Montgomeryshire, and the Wye in Radnorshire, with older records from Caernarfon and Meirionydd. The first record in England was from Slapton Ley, South Devon and the most recent record is also from still water, in 2009 in a quarry in Herefordshire. The last record in Ireland was in a ballast pit in North Kerry in 1929, earlier records being for rivers in South Kerry, Mid Cork and the vice-county of Dublin. The Guernsey record, in 1932, was also from a quarry. Recorded from April to October, peaking in July.

2. *Bidessus unistriatus* (Goeze) Plate 94

Length 1.7-2.0 mm. Females have heavy microreticulation on the upper side so they appear dull whereas the males are shiny. Found on soft substrata, mainly clay and peat, typically in very shallow water, in lowland ponds and ditches with fluctuating margins. Originally known in old fenland areas in England from Dorset to Norfolk, but now exceptionally rare, being confined to the New Forest and the Norfolk Broads. Recorded from February to November, peaking in April and August.

13. *HYDROGLYPHUS* Motschulsky

1. *Hydroglyphus geminus* (Fabricius) Plate 95

Length 1.9-2.2 mm. The typical habitat is still lowland waters with a disturbed and exposed substratum of clay, but *H. geminus* is also typical of the rhynes, shallow mossy ditches of the Somerset Levels and other places with fluctuating margins and silt. Found over much of England including Yorkshire, only recently recorded in Westmorland and County Durham the northernmost record being for South Northumberland in the 19th Century. The Welsh distribution is scattered including Anglesey, Breconshire, Caerfyrddyn, Denbighshire, Monmouthshire, Montgomeryshire, Pembrokeshire, and Radnorshire. *H. geminus* is also known from Guernsey and Jersey. Recorded throughout the year, peaking in April and September.

14. *DERONECTES* Sharp

1. *Deronectes latus* (Stephens) Plate 96

Length 4.2-4.8mm. This is a brown or reddish species, occasionally with a paler bar across the front of the elytra. It has been confused in the field with *Hydroporus rufifrons*, a beetle unlikely to be found in the turbulent habitats it occupies adjacent to fast running water, on exposed rock or among willow moss (*Fontinalis antipyretica* Hedw.) growing on that rock, often in shade and beneath undercut banks. In the north and west of Scotland it is also found in gravel beds in rivers and streams. In southern England *D. latus* is known recently from East Sussex, Surrey, North Hants and the New Forest, Dorset and is frequent in the south-west. It is absent from most of central England, with modern records for Derbyshire, Yorkshire, Cumberland, but not within the Lake District, and South Northumberland. The Scottish distribution is also patchy, mainly in headwater systems in the south, and then along the west coast from Ayrshire and Kintyre to Inverness-shire, with other modern records for Caithness and Moray. In Wales it would appear to have been known more frequently in the south but there are only modern records for Caerfyrddyn, Ceredigion, Caernarfon and Meirionydd. Island records are for Skye, Arran, the Isle of Man and the Isle of Wight. Recorded in January and from March to October, peaking in June and September.

15. *GRAPTODYTES* Seidlitz

These are small diving beetles, mainly black with yellow markings. Males have the fore and mid tarsi only slightly enlarged compared with females, but, apart from *G. flavipes*, they have other sexually related features, the inner fore claws being modified and the hind tibiae expanded.

Key 14. The species of *Graptodytes*

1. Elytra with distinctive pattern of a sharply defined black spot on a yellow background; head mainly red .... 4. *Graptodytes pictus* (Fabricius) (p. 72)

- Elytral pattern made up of lines, sometimes indistinct; head black 2

2. Elytra each with about 5 yellow lines .... 2. *Graptodytes flavipes* (Olivier) (p. 71)

- Elytra each with at most 2 lighter lines visible 3

3. Antennal segments 5-9 slender, longer than broad (Fig. 186); male front tarsal claws very unequal in length (Fig. 187); body almost parallel-sided .... 1. *Graptodytes bilineatus* (Sturm) (p. 71)

- Antennal segments 5-9 hardly longer than their width across the ends (Fig. 188); male front tarsal claws almost equal in length (Fig. 189); body more rounded ....
.... 3. *Graptodytes granularis* (Linnaeus) (p. 72)

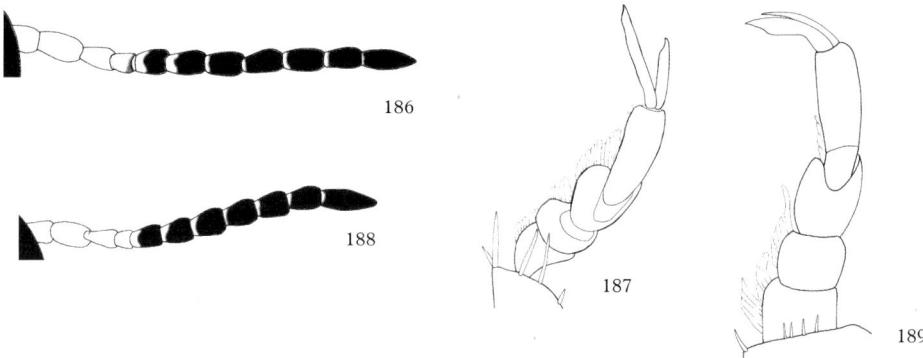

1. *Graptodytes bilineatus* (Sturm) Plate 97

Length 2.3-2.7 mm. This species is longer and more nearly parallel-sided than *G. granularis*; its longitudinal stripes are also more clearly marked. *G. bilineatus* occurs in England mainly in reedbeds, sometimes in brackish water. On the Severn it is currently known from the inner levels in West Gloucestershire though there are old records for North and South Somerset, and in Dorset it occurs on Cogden Beach. Otherwise it is eastern, being found on Humberside in South-east Yorkshire, around the Wash in South Lincolnshire and West Norfolk and inland to Northamptonshire, and then more widely in the Home Counties. The sole Welsh record is for Skomer Island in 1966, and it is also known from Jersey. In Ireland this species is confined to the karstic area of the west, mainly in turloughs and other lakes with vegetated fluctuating margins, in West Cork, Clare, all three vice-counties of Galway, Longford, Roscommon, and East Mayo. Recorded throughout the year, peaking in May and August.

2. *Graptodytes flavipes* (Olivier) Plate 98

Length 2.4-2.7 mm. *G. flavipes* is recognisable in the field as a small dark species with multiply striped elytra. It has simple male fore claws and a narrow symmetrical aedeagus whereas the other *Graptodytes* species have the inner fore claws lobed and the aedeagus blunt and asymmetrical. However, it should not be necessary to dissect specimens to confirm their identity. Typical of sparsely vegetated, hard-bottomed pools on acid rock and peat in heathland areas, also on clay. Currently confined to the Lizard, some Dorset Heaths and the New Forest It was formerly known from some intervening sites, from Pembrokeshire, including Skomer Island, and the Home Counties, the most outlying record being for Burwell Fen in Cambridgeshire. Also known from Guernsey. Recorded throughout the year, peaking in April and September.

3. *Graptodytes granularis* (**Linnaeus**) Plate 99

Length 2.0-2.3 mm. *G. granularis* is the commonest very small diving beetle in Britain and Ireland, a dark species with what is usually a poorly defined longitudinal stripe on each elytron. Associated with well vegetated, permanent ponds and ditches, often with fluctuating margins. Taking all records into account, *G. granularis* is scattered across the whole of lowland England, but it is frequent only from the east of Yorkshire to East Anglia, and in the New Forest and on Dorset heaths. There are then remnant populations on the edges of the Lake District, at Moccas Park in Herefordshire, along the Severn in Worcestershire, and around London. It is also scattered in Wales, being frequent only on Anglesey. In Scotland *G. granularis* is confined to the south-west coast in Wigtownshire and Kirkcudbrightshire. It is widely distributed and often common in Ireland. There are old records for Jersey. Recorded throughout the year, peaking in April and August.

4. *Graptodytes pictus* (**Fabricius**) Plate 100

Length 2.0-2.4 mm. This brightly coloured beetle has a similar shape to *Hygrotus decoratus* and the markings more sharply defined. *Hydrovatus cuspidatus* is also similar in markings and size. Any final doubts can be removed by examining the last visible abdominal sternite of the male, which uniquely has two yellow tufts of hair (Figure 190). Found in permanent ponds, lakes, canals and other slow-moving water with plenty of vegetation, often in enriched and man-made sites. The English distribution covers the whole of the lowlands apart from dry areas and the south-west. The more recent northern records are from along the Tyne and in old limestone workings along the coast stopping short of Scotland, where the species is mainly known in the south-west coastal vice-counties. However the species is also found spasmodically in the Central Belt, is common in Knapdale and ranges north to Arisaig in West Inverness-shire. Island records are for Eilean Dubh Beag in the Firth of Lorn in the southern Hebrides, also for the Isle of Man and the Isle of Wight. In Wales *G. pictus* is common on Anglesey and frequent along the coast and scattered inland. *G. pictus* is common across Ireland and is known from Jersey. Recorded throughout the year, peaking in June and August.

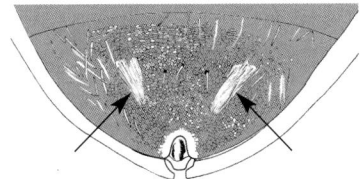

Figure 190. *Graptodytes pictus*, yellow tufts of hair on last visible abdominal sternite

16. *HYDROPORUS* Clairville

This is the most speciose genus of the British and Irish Dytiscidae and the most demanding in terms of identification. These are all small diving beetles ranging from 1.9 to 5.3 mm long; exact size is often crucial in determining the species involved.

Male genitalia

Until one is familiar with the slight differences in shape and size of many species of *Hydroporus*, and with the amount of variation, positive identification is best confirmed by examination of the male genitalia. An aedeagus, being strongly curved, can appear foreshortened and the structure will need to be tilted back and forth. The aedeagi of all species are depicted at their maximum possible length viewed dorsally. The side view of the aedeagus is also often useful, but the parameres of the male and the female genitalia are less useful diagnostically.

Sexing

Hydroporus can be difficult to sex because both sexes have sucker hairs on the fore and middle tarsi and, in many species, little difference in the width of these tarsi or the development of their claws. However, in some species the males have a lobe or tooth on one or both of the fore tarsal claws. Some species have dimorphic females, i.e. one form that resembles the males in the extent of microreticulation and another that has the microreticulation so intense that the insect appears quite matt. Thus extremely matt *Hydroporus* tend to be female and their presence can indicate the species involved. In one case at least, *H. memnonius*, the two forms have quite different distributions, the matt form *castaneus* Aubé appearing to be spreading at the expense of the type form; coupled with this it is possible to recognise the male of each form, as may ultimately prove the case for other dimorphic species.

Microreticulation

The beetle should be examined with the light from the side, at x 5 or more and when dry. Where microreticulation is present, the elytra have a slightly frosted appearance at x 12, at x 25 the net-like reticulation between the punctures is usually visible and at x 40 and above, the individual meshes are very obvious and can be counted, as may sometimes be required. Where there is no microreticulation, the pronotum, elytra and the last abdominal sternite are smooth and shiny, although dense hairs can make this less obvious at low magnification.

Colours

Although colour is important there will always be difficulties because of variation in the commonest species, *Hydroporus palustris*, which exists as the typical yellow chevron-patterned form and both as an almost entirely black form and as an orange one. Some species have shoulder markings, the "shoulders" being the front outer corners of the elytra, and the "shoulder bar" a pale mark extending across the front of the elytra. The key relies heavily on colour as *Hydroporus* fall into four colour-based groups, those that are entirely black, those that are a uniform brown, those with a contrast between a black pronotum and brown elytra, and those with distinct elytral patterns. Specimens preserved in alcohol will appear to be dark unless allowed to dry for 5-10 minutes. Dry specimens may need to be degreased by washing them with alcohol. Dry, stored specimens can fade, as do old beetles (these will also tend to have lost their hairs). Also, newly-emerged ("teneral") specimens will not show characteristic colouration; these can be recognised by the ease with which the cuticle is dented, and are best identified by reference to mature specimens. If there is doubt about colours, lift an elytron to see if a pattern is then visible. Finally, it should be noted that a tendency for beetles to become wholly dark, melanism, occurs in peaty areas in many species and that the opposite, normally dark parts being red – or rufinism – occurs in base-poor waters in some specimens of *H. melanarius*, *H. palustris* and *H. pubescens*. Such variants defy almost any form of key; it is advised that whenever possible *Hydroporus* species should be identified from a series in which any variation in colour can be set aside when all specimens in a series appear to be structurally identical.

Antennae

Count the eleven segments working from the head outwards. The outer halves of segments 4 onwards are often darkened, the almost entirely pale antennae of a few species setting them apart from the majority. Antennal segments vary in length from species to species but quantification of these differences can prove unsatisfactory as an identification aid initially.

The prosternal process
The neck of the process must be viewed on the underside of a clean dry beetle, looking between the fore coxae with the beetle's head tilted up at an angle of 45° and strongly illuminated.

Metacoxal processes
The rear edge of these can be diagnostic but difficult to see without manipulation of the light source to produce a highlight along the edge. Also diagnostic can be the extent of hairiness of these processes.

Shape
Slight differences in outline would appear to be something that the human eye can detect easily but measurements of which often fail to convince. Far from being optical illusions differences in features such as body outline are there to be seen and ultimately used to determine species – but they have little place in these initial attempts at identification, hence restricting their mention in the key, which concentrates on size, colour and male genitalia. Plates 101-129 are recommended as a guide to body shape.

Key 15. The species of *Hydroporus*

1. Elytra with patterning or pale areas at front contrasting with a dark pronotum (for example Plates 117 and 121) 2

- Elytra and pronotum similarly coloured where they meet (for example Plates 102 and 107) 16

2. Centre of pronotum and large parts of elytra devoid of microreticulation 3

- Centre of pronotum and elytra microreticulate 5

3. Pronotum black with at most the lateral raised rim of the margins slightly paler; metacoxal processes shining with a thin covering of hair; pronotum and elytra shining; aedeagi without a distinct point (Figs 193-195) 4

- Pronotum with broad pale margins; rear half of metacoxal processes completely covered with hair, sockets of which are visible at an intense density if the hairs have been abraded (Fig. 191); punctures so dense on pronotum and elytra that they appear matt; aedeagus bluntly pointed (Fig. 192) .... 11. *Hydroporus marginatus* (Duftschmid) (p. 85)

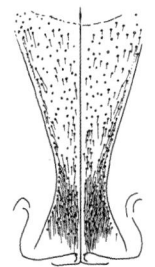

191

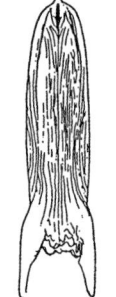

192

4. Length 3.2-3.8 mm; last abdominal segment without microreticulation, shining between punctures; aedeagus (Fig. 193) .... 22. *Hydroporus pubescens* (Gyllenhal) (p. 88)

- Length 3.8-4.7 mm; last abdominal segment microreticulate; aedeagus (Fig. 194) .... 21. *Hydroporus planus* (Fabricius) (p. 88)

193   194

5. Whole of head and pronotum black; elytra in contrast often with a grey pattern of many streaks or mottles; body distinctly shining as the microreticulation partly effaced; length 3.4-3.8 mm; aedeagus blunt (Fig. 195) .... 26. *Hydroporus tessellatus* (Drapiez) (p. 89)

- Head red or brown at front, paler than pronotum; elytral pattern, if present, either one based on distinct yellowish markings or one a dark longitudinal mark around the suture; microreticulation well developed; aedeagus weakly or sharply pointed 6

195

6. Outer antennal segments darkened (Fig. 196); elytra often patterned but not semitransparent such that the tracheae of the elytra and parts of the underlying wings can be seen 7

196

- Antennae generally pale, terminal segment sometimes slightly darkened; elytra semitransparent 15

7. Pronotum with sides converging at the rear or parallel such that the body outline is broken up when viewed from above (Plate 117); inner fore claw of male with a distinct lobe (Fig. 197); aedeagus distinctive (Fig. 198); not exceeding 2.7 mm in length .... 16. *Hydroporus neglectus* Schaum (p. 86)

- Pronotum with sides diverging to make a more or less continuous outline with the elytra (e.g. Plate 104); inner fore claw of male thickened in one species, otherwise the same as the other claw; aedeagi not as Fig. 198; one species down to 2.5 mm in length, otherwise all larger than *H. neglectus* 8

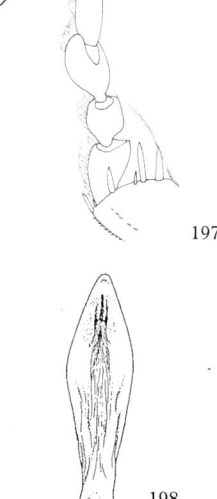

197

198

8. Neck of prosternal process with a distinct step but without grooves (Fig. 199); elytra black, usually with distinctive yellow pattern of chevrons that may be reduced to marks along the sides; male with inner fore tarsal claws more strongly curved and thicker than the outer ones (Fig. 200); aedeagus shaped like a tapering sword (Fig. 201); length 3.3-4.0 mm ....
.... 20. *Hydroporus palustris* (Linnaeus) (p. 88)

- Neck of prosternal process with grooves (Fig. 202); either elytra a contrasting brown to the black of the pronotum or very dark with shoulder bar or obscure lateral marks; male fore tarsal claws alike; aedeagi not as figured for *H. palustris* 9

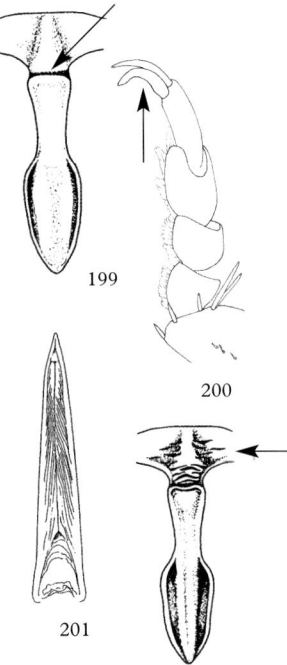

199

200

201

202

9. Over 3.4 mm long; lateral keel of elytra, when viewed from the side, almost straight, making an angle with the pronotum greater than 140° (Fig. 203); aedeagi broad to near the point (Figs 205 and 207) 10

- If over 3 mm then lateral keel curved, making a stronger but still obtuse angle (less than 130°) with the pronotum (Fig. 204); aedeagi narrow or narrowed from a broad base (Figs 206 and 209-211) 12

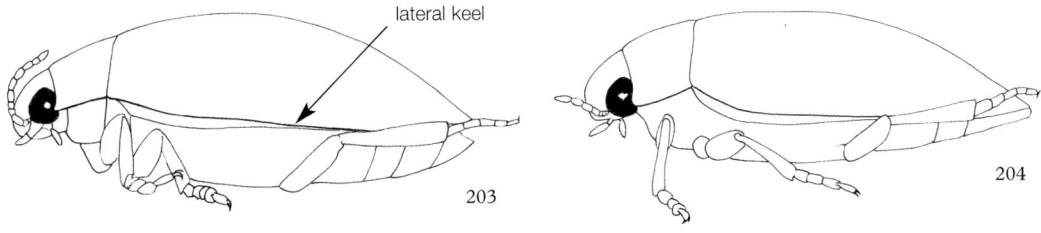

lateral keel

203

204

10. Length 4.0-5.3 mm; pronotal rim wider than the maximum thickness of a fore claw; females with the same microreticulation as the males; aedeagus large and broad, asymmetrical with greatly expanded sides (Fig. 205) ....
.... 23. *Hydroporus rufifrons* (Müller) (p. 89)

- Length 3.3-4.5 mm; pronotal rim narrower than the maximum thickness of a fore claw; females sometimes more intensely microreticulate than males; aedeagus broad and drawn out into a blunt or elongate point 11

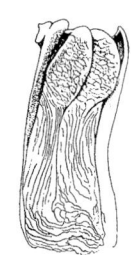

205

11. Length 3.4-4.5 mm; an imaginary line drawn following the pronotal edge would smoothly connect to a line following the elytra; elytra widest at about their mid-point (Plate 104); hind edge of last abdominal sternite crammed with punctures less than a puncture's diameter apart; females often completely dull (var. *deplanatus* Gyllenhal); aedeagus broad with a small finger-like process (Fig. 206) ....
.... 4. *Hydroporus erythrocephalus* (Linnaeus) (p. 83)

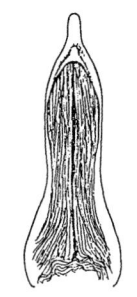

206

- Length 3.3-3.8 mm; that imaginary line would demonstrate that the rear edge of the pronotum is slightly wider than the elytra; elytra slightly flared so that they are widest about ⅔ rds along; hind edge of last abdominal sternite with widely spaced punctures; females with microreticulation the same as in the males; aedeagus broad with a blunt point (Fig. 207) ....
.... 3. *Hydroporus elongatulus* Sturm (p. 83)

207

The slight expansion of the rear edge of the pronotum provides a good character best observed when this very rare species is compared with the other species from couplet 9 to this point.

12. Pronotum black or brown, sometimes with paler sides, elytra with a pale band at the shoulders, usually with epipleurs partly pale (Fig. 208); aedeagi constricted at tip to form a short process or at least a blunt point (Figs 209-211) 13

If in doubt lift an elytron to check.

208

- Pronotum entirely black, elytra brown sometimes with a dark longitudinal mark either side of the suture but never with a pale shoulder band 14

13. Ground colour dark brown or black with diffuse lateral streaks on elytra sometimes extending as a greyish patch nearer the middle at the front; ♂ last abdominal sternite with punctures spread out amongst the microreticulation; aedeagus (Fig. 209); length 2.9-3.4 mm .... 25. *Hydroporus striola* (Gyllenhal) (p. 89)

209

- Ground colour brown with a cream or yellow patch near the middle at the front of each elytron, also lateral streaks of the same colour; ♂ last abdominal sternite with dense puncturation reducing the amount of microreticulation; aedeagus (Fig. 210); length 3.3-4.0 mm .... 8. *Hydroporus incognitus* Sharp (p. 84)

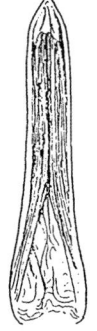

210

14. Edges of pronotum distinctly raised into a rim about as wide as the trough running alongside its inner edge; head appearing wide, the maximum width of the pronotum is not more than 2.21 times the maximum width of the head; body more parallel-sided – see Plate 126; brightly coloured with sparse hairs; aedeagus narrowly rounded at tip (Fig. 211) ... 27. *Hydroporus tristis* (Paykull) (p. 90)

- Rim of pronotum very fine, less wide than one of the coarse leg bristles and not raised; head appearing narrower than in *H. tristis*, the maximum width of the pronotum not less than 2.25 times the maximum width of the head between the eyes; body shape more rounded – see Plate 127; dull with much hair, though sometimes abraded; aedeagus broadly rounded at tip (Fig. 212) ....
.... 28. *Hydroporus umbrosus* (Gyllenhal) (p. 90)

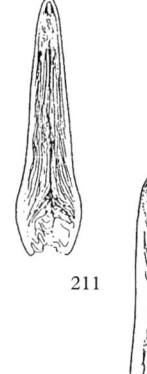

211

212

15. Middle of elytra with very small punctures (visible at x 40) in between sparse large punctures (of at least two types) more than twice their own diameter apart (Fig. 213); pronotum with a group of heavy elongate punctures on the hind margin near the sides and a row along the front edge, but almost no large punctures elsewhere; prosternal process without grooves; aegeagus (Fig. 214); length 3.3-4.2 mm ....
.... 19. *Hydroporus obsoletus* Aubé (p. 87)

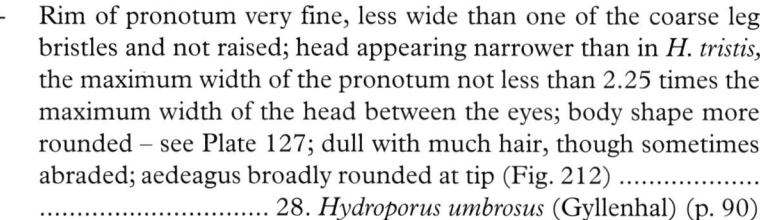

213

The small punctures, which can be found on other parts of the body, are unusual among *Hydroporus* and, in the British and Irish faunas, are only found in any number in this species and *H. memnonius*.

- Centre of elytra without much smaller punctures in between large punctures less than twice their own diameter apart; pronotum with large punctures more widely distributed, though with a concentration at the hind margin and fewer in the centre; prosternal process with grooves; aedeagus (Fig. 215); length 3.5-4.2 mm .... 5. *Hydroporus ferrugineus* Stephens (p. 83)

214

215

16. At least part of the hind half of the pronotum and most of the elytra devoid of microreticulation 17

- Whole of pronotum and elytra microreticulate 18

17. Body wholly black, pronotal rim sometimes a little lighter; last abdominal sternite microreticulate with the ♂ sternite densely punctured; aedeagus with sides only narrowly visible in dorsal view (Fig. 216); length 3.0-3.5 mm ....
... 2. *Hydroporus discretus* Fairmaire & Brisout de Barneville (p. 82)

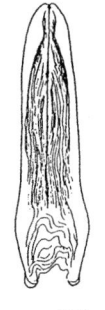

- Most of body black but pronotal margins and elytra a little lighter; last abdominal sternite (in both sexes) shining, devoid of microreticulation and with widely spaced punctures; aedeagus with sides widely visible in their entirety in dorsal view (Fig. 217); length 3.1-3.8 mm ... 22. *Hydroporus pubescens* (Gyllenhal) (p. 88)

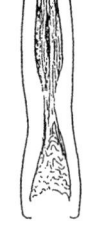

216

This couplet is to catch darker *pubescens* (see couplet 4): the body shapes here (*discretus* a broad oval, *pubescens* more elongate oval) are confirmatory but of little value if few specimens are available for comparison. *H. pubescens* is the only British species with a strongly shining last abdominal sternite.

217

18. Upper side almost entirely black, though sides may be dark brown 19

- Upper side predominantly brown or red 27

19. Rear edge of each metacoxal process scalloped to produce a wavy margin overall (Fig. 218); aedeagus viewed from above asymmetrical (Fig. 219); antennae pale ....
.... 10. *Hydroporus longulus* Mulsant & Rey (p. 84)

- Rear edges of metacoxal processes straight (Fig. 220) or, if slightly curved, then they protrude further in the midline than at the outside edges (Fig. 221); aedeagus symmetrical; antennal segments darkened or pale 20

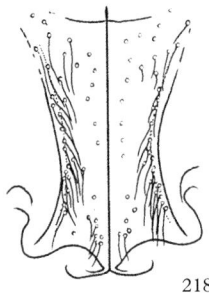

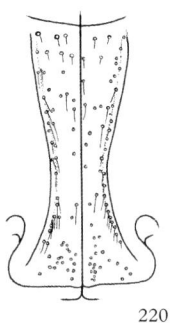
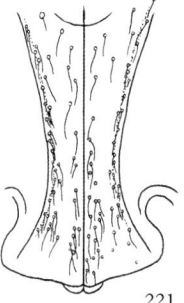

218

219

220

221

20. Tarsi and lower parts of tibiae black viewed from above 21

- Legs mainly reddish or yellowish with tarsi only partly, if at all, darkened 22

Dark *H. memnonius* and *H. gyllenhalii* could key to this couplet; if length over 3.5 mm, see couplet 30.

21. Head all black beneath; male with inner fore claw toothed (Fig. 222); females slightly duller than males; aedeagus broad (Fig. 223) without a deflected tip; length 3.3-3.5 mm .... 14. *Hydroporus morio* Aubé (p. 86)

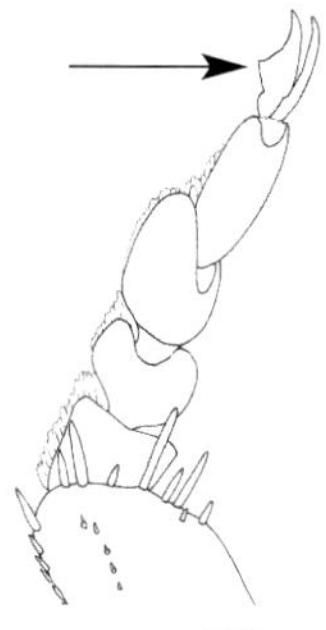

- Head with paler markings beneath; inner fore claw of male without a tooth; aedeagus narrow and appearing blunt in dorsal view (Fig. 224) and with a downwardly pointing projection in side view (Fig. 225); females distinctly duller than males; length 2.8-3.2 mm .... 6. *Hydroporus glabriusculus* Aubé (p. 83)

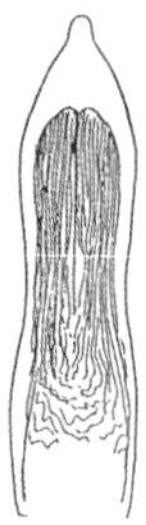

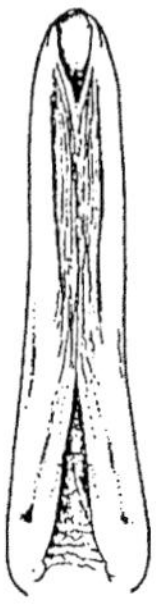

22. Large (3.8-4.5 mm) flat insect; metacoxal processes with few punctures or hairs on the area around the midline; antennal segments long, most segments at least twice as long as wide; aedeagus with a tongue-like process (Fig. 226) .... 13. *Hydroporus memnonius* Nicolai (p. 85)

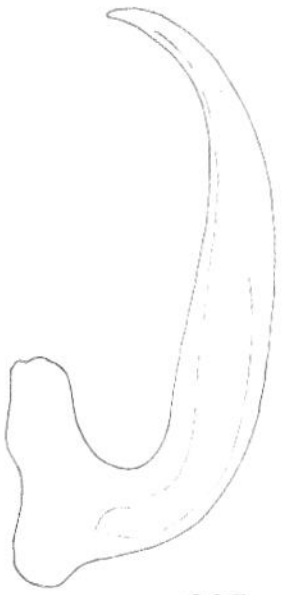

continue to couplet 23

- 3.8 mm long or less; aedeagus without a tongue-like process; not with more than two of the other characters above 24

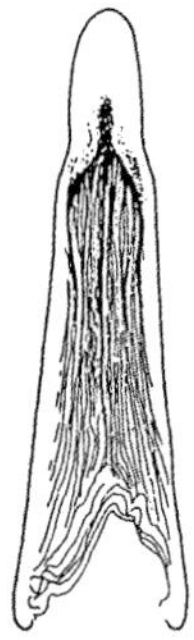

23. Males and females both shining .... 13. *Hydroporus memnonius* type form (p. 85)

- Female intensely microreticulate .... 13. *Hydroporus memnonius castaneus* Aubé (p. 85)

The male of this form has almost twice as many large suckers (14-18) on the first tarsal segment of the fore legs as in males from the all shining population (8-15) and two large suckers on the second tarsal segments of the fore and middle legs whereas the all shining population has none.

24. Antennae appearing short, with middle segments slightly wider in the male than in the female, segments 5-11 strongly blackened (Fig. 227); metacoxal processes only sparsely hairy; aedeagus blunt (Fig. 228); a small insect (2.8-3.5 mm) slightly domed and with a rounded body outline ....
.... 17. *Hydroporus nigrita* (Fabricius) (p. 87)

227

\- Antennal segments weakly if at all darkened and appearing longer without sexual differences; metacoxal processes with a weft of long hairs centrally (Fig. 229), if abraded then marked by dense puncturation; aedeagi pointed; generally larger (3.0-3.8 mm) than *nigrita*, flatter and with the elytral sides straighter so that body does not appear rounded 25

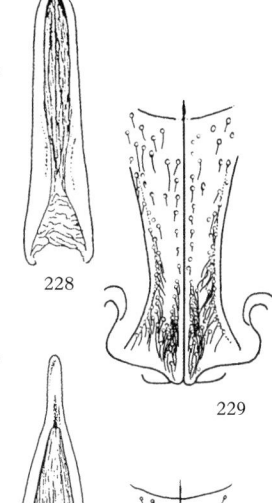

228

229

25. Aedeagus with long process at its tip (Fig. 230); rear edges of metacoxal process with inner corners drawn out so that the edges have a sinuate appearance (Fig. 229); length 3.5-3.8 mm; body shape narrow and parallel-sided (Plate 109) ....
.... 9. *Hydroporus longicornis* Sharp (p. 84)

\- Aedeagus with a wide blade at its tip (Figs 232 and 233); rear edge of metacoxal processes straighter (Fig. 231); length not exceeding 3.6 mm; body shape broader, either parallel-sided (Plate 112) or rounded (Plate 116) 26

The two following taxa can only be separated reliably using male genital characters.

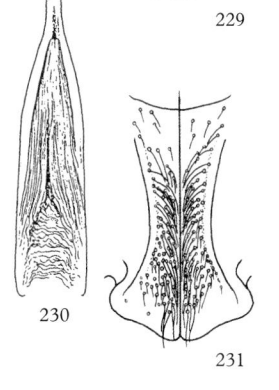

230

231

26. Aedeagus with the end part (the section beyond the sperm duct opening, which lies in front of the wrinkled part of the dorsal surface) short, 0.07 mm long (Fig. 232); elytra largely straight-sided, giving a parallel-sided appearance (Plate 112); length 3.0-3.5 mm .... 12. *Hydroporus melanarius* Sturm (p. 85)

\- Aedeagus with longer end part, about 0.11 mm long (Fig. 233); elytra continuously rounded in profile (plate 116); length 3.2-3.5 mm .... 15. *Hydroporus necopinatus* Fery (p. 86)

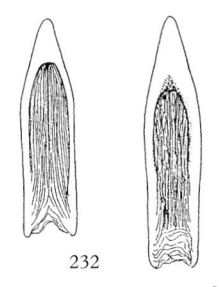

232

233

27. In dorsal view the pronotum meets the elytra at an angle breaking up the outline of the body (Plates 101 and 125); males with lobes below the tips of the fore claws (Fig. 234) 28

\- Pronotum and elytra forming a smooth outline to the body; fore claws without lobes 29

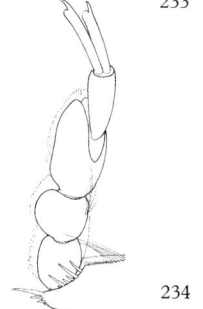

234

28. Length 2.8-3.3 mm; aedeagus with a flat asymmetrical process (Fig. 235) .... 1. *Hydroporus angustatus* Sturm (p. 82)

- Length 1.9-2.2 mm; aedeagus symmetrical with a long terminal process (Fig. 236) ..... 24. *Hydroporus scalesianus* Stephens (p. 89)

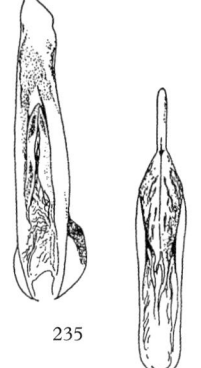

235

236

29. Length not more than 3.1 mm; centre of pronotum with only very fine punctures in contrast to the elytra; no longitudinal rows of punctures visible on elytra; aedeagus (Fig. 237) .... 18. *Hydroporus obscurus* Sturm (p. 87)

- Length 3.4-4.3 mm; centre of pronotum with some large punctures; only a little less marked than on elytra; elytra with or without lines of punctures 30

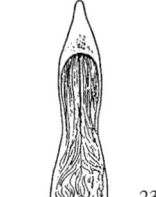

237

30. Punctures of two sizes, the larger ones outnumbering the smaller ones, also punctures in well defined rows – or with only small punctures in intensely microreticulate and matt ♀; body outline partly parallel-sided; aedeagus with a tongue-like process (Fig. 238); length 3.8-4.3 mm ...... 13. *Hydroporus memnonius* Nicolai (p. 85)

Return to couplet 23 for more information.

- Punctures large and distinctively deep; elytral outline more rounded and usually without rows of punctures; aedeagus pointed (Fig. 239); length 3.4-4.1 mm .... 7. *Hydroporus gyllenhalii* Schiødte (p. 84)

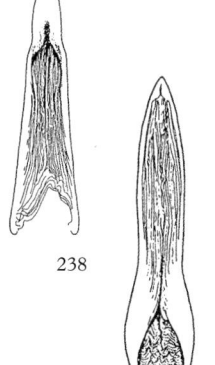

238

239

1. *Hydroporus angustatus* **Sturm** Plate 101

Length 2.8-3.2 mm. A narrow insect, reddish particularly in its front half. All the fore claws are elongated and lobed in the male, and the male hind tibiae have a distinct kink. Associated with permanently flooded fens, usually in mesotrophic but also in enriched sites. Common across mainland, lowland Britain and Ireland in all but the drier areas north to East Ross. Largely absent from mountainous areas. Known with certainty from Inishbofin island in West Galway, the Isle of Man, and the Isle of Wight, Guernsey and Jersey, but requiring confirmation from Skye and Shetland. Recorded in all months, peaking in May and September.

2. *Hydroporus discretus* **Fairmaire & Brisout de Barneville** Plate 102

Length 3.0-3.4 mm. A broadly oval species most likely to be confused with *H. nigrita* because of its shape or *H. pubescens* because of its hairiness and lack of microreticulation.

Mainly found in muddy headwaters and springs. Ranging across the whole of Britain and Ireland north to the Orkneys, frequent in uplands and much of the Hebrides, scarce or absent from fenland areas. Recorded from Guernsey, Jersey and Lundy. Recorded in all months, peaking in June.

3. *Hydroporus elongatulus* **Sturm** Plate 103

Length 3.3-3.6 mm. In the field this species resembles *H. gyllenhalii* but it is flatter with the slightly flared elytra making an angle with the pronotum. *H. elongatulus* is confined to mesotrophic fen, usually in shallow water with rushes, sedges and moss. The oldest record is for Roxburghshire in 1956 but it was not detected until 1974, since when it has been recorded from Ayrshire, Berwickshire, Dumfriesshire, and Peeblesshire, and in Norfolk and Suffolk on the Brecks, mainly in "pingo fens". Recorded in all months except October, peaking in May and August.

4. *Hydroporus erythrocephalus* **(Linnaeus)** Plate 104

Length 3.4-4.3 mm. A highly arched species that usually has a strong contrast between the red ground colour of the head and the darker pronotum. The elytral shoulder may be a pale yellow, making this species look like a large *H. incognitus*. Frequent in a wide range of permanent, still water habitats to about 500 metres above sea level across the whole of Britain and Ireland to the extreme north of the Shetlands and reaching islands from Guernsey, Jersey and the Scillies to Tory Island and remote islands such as Fair Isle and Foula. This species is dimorphic in its form of female, both male-like and matt females (*deplanatus*) occurring together, though with the matt form commoner on high ground and in the north. Recorded in all months, peaking in June and September.

5. *Hydroporus ferrugineus* **Stephens** Plate 105

Length 3.5-4.2 mm. This flat species is brown with a mottled pattern, likely to be confused with the less clearly marked *H. obsoletus*. It is truly subterranean, found deep in caverns in Derbyshire, but more usually found at the mouths of springs after heavy rain. *H. ferrugineus* ranges across the whole of Britain, but is frequent only in the Central Weald, in the Pennines and in Snowdonia. Despite its apparent flightlessness *H. ferrugineus* is known from several islands, Anglesey, Bute, Mull and, requiring confirmation, Hirta in the St Kilda complex. It is known high on mountains, such as Cairn Gorm, but has also been pumped up in well water in the Cambridgeshire Fens and in the centre of Glasgow. Recorded in all months, peaking in June and October.

6. *Hydroporus glabriusculus* **Aubé** Plate 106

Length 2.8-3.2 mm. A narrow blackish insect intermediate in appearance between *H. striola* and *H. morio*. The female is always matt, so the finding of such dull, black *Hydroporus* should alert one to the presence of this very rare species. *H. glabriusculus* is typical of deep, relict fen, particularly where the surface structure is broken up into a series of small pools as holes in quaking bogs or amongst tussocks, and it is often abundant. The first British record was from Askham Bog in 1906, where the species was originally identified as *H. morio*, but it is no longer known from Yorkshire, modern records being in clusters in Berwickshire, Roxburghshire and Selkirkshire in the Scottish Borders and in the Brecks in the West Norfolk "pingo fens" extending east to two sites away from the Broads in East Norfolk, and at one point within Catfield Fen in the Broads proper. It may prove more frequent in the centre of Ireland with records from Cavan, Fermanagh, Sligo and Westmeath. Recorded in all months, peaking in April and August.

7. *Hydroporus gyllenhalii* Schiødte Plate 107

Length 3.4-4.1 mm. This species causes many problems for beginners, often confused with other *Hydroporus* in its size range: the coarse punctuation is its most distinguishing feature, being otherwise a rather nondescript species. It is often common in dystrophic or mesotrophic pools, but it occasionally occurs in small numbers in circumneutral waters. Frequent across northern Britain north of a line from the Severn to the Humber, also in south-west England and in a broad band from Dorset to north of London, in the Weald and in East Anglia, but with large gaps on the east coast and in dry areas. *H. gyllenhalii* penetrates to many remote islands including St. Kilda, Fair Isle, Lundy and the Scillies. It is common in Ireland, even in karstic areas where there is surface peat. Recorded in all months, peaking in April and September.

8. *Hydroporus incognitus* Sharp Plate 108

Length 2.9-3.9 mm. *H. incognitus* is similar in size to *H. palustris* and often occurs with it, adding to the confusion; in its most distinctive form it is a shiny dark brown with a cream or yellow bar at the elytral shoulder – but dark, duller and less distinctly marked specimens do occur. The typical habitat is small shaded pools amongst dead leaves, but in the north-west of Scotland the range is extended by its living in saltpans at the landward edge of saltmarshes, usually full of rotting seaweed. Although this species is frequent across much of Ireland and Britain north to the Orkneys there are distinct gaps in the distribution, particularly in inland England south of Leicestershire and in the Highlands. Recorded in all months, with peaks in May and October.

9. *Hydroporus longicornis* Sharp Plate 109

Length 3.5-3.8 mm. *H. longicornis* is the most parallel-sided *Hydroporus*, narrower than *H. melanarius* and with slightly darkened antennae unlike the clear antennae of *H. longulus*, the other species with which it can be confused. Always associated with slow flowing water in dense mossy vegetation, either in headwater seepages or in valley fens, in the south-east of England confined to areas with woodland cover. In England *H. longicornis* is found along the Pennines, in the North Yorkshire Moors and in the Lake District, mainly in headwaters, and then scattered in isolated valley fens and bogs in southern England in East Cornwall, South and North Devon, South Somerset, Dorset, South Hampshire, East Sussex, East Kent, Buckinghamshire, and East and West Norfolk. It is also frequent in North and central Wales. The Scottish distribution is mainly western, ranging from Wigtownshire to Caithness. There are comparatively few records for Ireland, in South Kerry, Waterford, Wicklow, Limerick, Sligo, West Donegal and Fermanagh. Island records are for Arran, the Holy Isle off Arran, Islay, Jura, Rum, Skye, and the Isle of Man. Recorded in all months, being frequent in the winter, peaking in May and September.

10. *Hydroporus longulus* Mulsant & Rey Plate 110

Length 3.4-3.8 mm. This shining black insect is quite distinctive, with yellow antennae and wavy margins to rear edges of the metacoxal processes. Confined to running water, usually in runnels on silt or rock, and in fissures in small springs. It is frequent in northern England in the Lake District and North Yorkshire Moors and along the Pennine chain extending to woodland in Leicestershire. The southern England distribution is scattered but with many records along the south-west coast and in the Weald and adjacent areas in Surrey, North Hampshire and Berkshire: *H. longulus* is almost entirely absent from East Anglia, there being only two modern records, from Cambridgeshire and West Suffolk. It is widely distributed in Wales except in the south and it also occurs in neighbouring hill land in England in Herefordshire and Shropshire. Scottish records are almost associated with hill land particularly in the Southern Uplands and the Eastern Highlands. *H. longulus* is found on the coastal mountain ranges all around Ireland. Island records include Anglesey, Arran, Clare Island, Harris and Lewis, Islay, Little Cumbrae, Lundy, the Isle of Man, Raasay, Rum, the Shetlands (but not the Orkneys), St. Kilda, Skokholm, Skomer, Skye, and South Uist. Recorded in all months, peaking in June.

11. *Hydroporus marginatus* (Duftschmid) Plate 111

Length 3.5-4.5 mm. *H. marginatus* resembles *H. planus* but has wide pale margins to the pronotum and is duller than *H. planus* owing to the more intense puncturation. Intermittent streams on chalk and limestone provide the habitat for this species, generally regarded as semisubterranean. However, *H. marginatus* is found in a much wider range of habitats in the south of Europe and it can occur in still waters in abandoned quarries in places such as the Cotswold Water Park. The modern distribution, based on records from Dorset, South and North Wiltshire, North Hampshire, East Kent, Hertfordshire, Berkshire, Oxfordshire, Buckinghamshire, and West Gloucestershire, is decidedly more compact than that indicated by old records. *H. marginatus* has only been recorded once in Wales, in the 19th Century in Glamorgan. Recorded in January and from March to November, peaking in May and September.

12. *Hydroporus melanarius* Sturm Plate 112

Length 3.0-3.6 mm. *H. melanarius* is black, flat and parallel-sided but rather squat compared to *H. longicornis*. Reddish specimens are usually teneral, but some populations have a high proportion of red-brown mature beetles. It occupies very shallow, temporary, still, acid water such as on the actively growing surface of raised bogs, or amongst tussocks in swamps, or occasionally in woodland pools. The distribution follows the availability of relatively undisturbed shallow acid waters across Britain. In England there is a large gap in the distribution from lower Humberside to Dorset, where the heathland pools are occupied by *H. necopinatus* Fery and *H. melanarius* is absent. In East Anglia records are mainly for the Broads and the Greensand of Norfolk. *H. melanarius* is frequent over much of Wales. It is scattered across most of Scotland north to the Orkneys. Other islands from which it has been recorded include Anglesey, Arran, the main Hebrides, and the Isle of Man. Recorded in all months, peaking in April and October.

13. *Hydroporus memnonius* Nicolai Plates 113 and 114

Length 3.8-4.3 mm. *H. memnonius* occurs not only as the matt and shining forms but also in black and chestnut brown colour forms. The matt pale brown *castaneus* females are often confused with other species, e.g. *H. ferrugineus*. Irrespective of form *H. memnonius* is associated with very shallow water in the edges of a wide range of water body types, though most typically amongst detritus in woodland pools and ditches. As with *Agabus uliginosus* the two forms of female of this taxon have distributions suggesting different routes or times of arrival in Britain. In this case it is also possible to differentiate the males, based on additional sucker hairs on the fore and mid tarsi of those associated with females of the matt form *castaneus*. Records for all *memnonius* cover the whole of Britain and Ireland to Shetland, reaching many remote islands but there are no records from Mull and large tracts of the Highlands. The matt form dominates in England and Wales, reaching the Scillies and Lundy, and extending into Scotland only up to a line from coastal Dumfriesshire to East Lothian. The shining form occurs across Ireland, Anglesey, the Isle of Man and Scotland, with small populations around the Lake District, and on the Lleyn Peninsula and in Pembrokeshire from St. David's Head to Fishguard. The two forms coexist in a few localities along the Scottish Borders and in the south of the Lake District, with some evidence of displacement of the shining form by *castaneus* in the past 30 years. Recorded in all months, *castaneus* peaking in May, the shining form in June, both peaking in September.

14. *Hydroporus morio* Aubé Plate 115

Length 3.0-3.7 mm. This black insect is recognised in the field by its pear-like shape, the male inner fore claw's tooth being diagnostic. The typical habitat is in shallow acid pools with a covering of matted dead grasses over peat, but this species can also be found in *Sphagnum* at the edges of larger pools. In England *H. morio* is mainly found on the Cheviots, Pennines and North Yorkshire Moors, and in the Lake District, with outlying records for Delamere Forest in Cheshire and Dersingham Bog in West Norfolk. *H. morio* ranges south in Wales to the Brecon Beacons; it is also known from Parys Mountain on Anglesey and Mynydd Ffoesidoes in Radnorshire, and on the mountains of Meirionydd and Caernarfon. It is found on all Scottish mountain systems, coming near to sea level in the Flow Country, where it has been found in a base-enriched swamp. The Irish distribution follows the coastal mountain systems, originally south to Mangerton Mountain in North Kerry and the Comeragh Mountains of Waterford, but modern records extend south only as far as the Wicklow Mountains. *H. morio* occurs on Snaefell in the Isle of Man and other islands with records include Arran, Eigg, Hoy, Islay, Jura, Mull, the Orkney Mainland, Raasay, Rum, Skye, South Uist, and Unst. Recorded in all months, peaking in June.

15. *Hydroporus necopinatus* Fery Plate 116

Length 3.2-3.5 mm. The shape of *H. necopinatus* should differentiate it from *H. melanarius*, and this is one case among *Hydroporus* where the geographical location, on Dorset heaths, is highly indicative. Its colour is black or dark brown. *H. necopinatus* occurs in small, hard-bottomed pools in peat around the edges of valley mires on exposed heathland. It is known from Dorset and the Channel Isles. The Dorset population has been described as ssp. *roni* Fery, distinct from ssp. *robertorum* Fery, found on Guernsey, Jersey and France, and the nominate form, *H. necopinatus necopinatus*, in northern Spain and in Portugal. Studies of mitochondrial DNA (David Bilton, pers. comm.) indicate a complex relationship between these forms and *H. melanarius*, supporting recognition of the endemic English form. Recorded from March to October, peaking in June and September.

16. *Hydroporus neglectus* Schaum Plate 117

Length 2.3-2.7 mm. This is a delicate, narrow insect, resembling a small *H. tristis* with which it is often found. It can also be found with *H. umbrosus*, a darker, more hairy insect of the same size. The male inner fore claw is toothed, a good character, and the aedeagus, with its downward deflection of the top combined with the middle part being expanded, is also distinctive. Associated with still water over peat or detritus such as the quaking edges of *Sphagnum* carpets growing in woodland pools and other acid or mesotrophic fens with mosses, also in woodland puddles with dead leaves. *H. neglectus* is largely confined to old fenland and heathland, sometimes in isolated, man-made pools such as in abandoned brickworkings. In England this species ranges north to Gormire Lake in North-east Yorkshire, south to the Thorne Moors. Other major clusters are in Cheshire, in East Anglia in the Brecks and Broads, on heathland from Dorset to the north of London, and in the Weald. In Wales the species is known from Radnorshire at Abercamlo Bog and from Denbighshire in Borras Bog, Wrexham, and in Fenn's Moss adjacent to Whixall Moss in Shropshire, where it can be abundant. For Ireland there is one record for a coniferous plantation beside Lough Graney in Clare. Known from Jersey. Recorded in all months, peaking in April and October.

17. *Hydroporus nigrita* (Fabricius) Plate 118

Length 2.8-3.3 mm. This is a squat black species dull because of the amount of microreticulation distinguishing it from *H. discretus* with which it often occurs. *H. nigrita* is common in seepages and puddles, usually with mud and grasses, ranging from coastal cliffs to mountain headwaters. Frequent across south-west England, Wales and most of northern Britain to the Shetlands, more patchy in the lowlands in the Midlands and south-east England. Reaching many small islands, such as Ailsa Craig in Ayrshire, Trondra in Shetland, Inishmaan in the Aran Islands and Saltee in Wexford. Recorded in all months, peaking in April and June.

18. *Hydroporus obscurus* Sturm Plate 119

Length 2.5-2.9 mm. The smooth body shape, broadest at the middle and tapering to the tip, coupled with the dominantly red colour of this small insect, will usually distinguish this species in the field. It is mainly associated with permanent water in *Sphagnum* bogs but also can be found along the exposed shores of base-poor lakes, some on marl deficient in phosphate. *H. obscurus* has a distinctly western bias to its distribution, common in the Western Isles, the west of Ireland, much of Wales and moorland and heathland in south-west England. In Scotland it extends to the north of the Shetlands and is common in lochs in the Cairngorms and in the Border Mosses as well as in the west. The English distribution is based on peatlands on moors and heaths, often on old commons, and it is very rare in East Anglia where it is largely confined to Catfield Fen and to the Greensand, but it is still frequent in the east on the North Yorkshire Moors and the Thorne Waste. Similarly, whilst this species is common in western Ireland, there are centres in the east, in particular the interdrumlin mires of Armagh and Down. Recorded in all months, peaking in June.

19. *Hydroporus obsoletus* Aubé Plate 120

Length 3.3-4.2 mm. *H. obsoletus* resembles a small version of the brown form of *H. memnonius* and it could be mistaken for a poorly marked *H. ferrugineus*. *H. obsoletus* is semisubterranean in the north of its range in Britain and Ireland, living in the twilight zone of springs and caves and being flushed out by heavy rain. The scattered distribution is mostly on base-poor rock, with an unusual feature being many records associated with coastal extremities. The northernmost records are not modern, from Unst in the Shetlands, Hoy in the Orkneys, and from Lewis. More recent northern records are from Raasay and Glen

Nevis. There are more records for southern Scotland down to its southern extremity on the Mull of Galloway, there matched by Langness on the Isle of Man and Holyhead Mountain on the Holy Island off Anglesey. Other island records are for Arran, Bute, Clare Island, Jura, and Ulva. Inland records are most frequent in southern Scotland and around the Lake District running down the Pennines, with records more scattered to West Cornwall and East Suffolk, the species being otherwise absent from south-east England. Irish records are for Antrim, Armagh, Clare, Down, Kildare, Limerick, Londonderry, North Kerry, South Tipperary, and West Mayo. *H. obsoletus* has occasionally been reported in company with *H. ferrugineus*, for example in springs around Loch Fad on Bute, in Hawthorn Dene in County Durham, and in the Carlswark Cavern in Derbyshire. Recorded in all months, with the greatest number of records from April to June and in August and September.

20. *Hydroporus palustris* (Linnaeus) Plate 121

Length 3.3-4.0 mm. Associated with almost any vegetated still or slow-moving water from coastal levels to montane lakes, also amongst plant debris in shaded habitats. Common across Britain and Ireland in all but the drier areas, found on some of the most remote islands, but not known from St. Kilda or the Scillies. *H. palustris* occurs on Alderney, Guernsey and Jersey. Colour may vary from almost entirely black, often in peaty areas, to entirely orange, as in some montane pools and in Loch Eck in Argyll. Frequent throughout the year, peaking in June and August.

21. *Hydroporus planus* (Fabricius) Plate 122

Length 3.8-4.8 mm. The size of *H. planus*, coupled with its hairiness, its black head and the grey colour of the front of the elytra, sets it apart from all species except possibly the rare *H. marginatus*, which has much denser puncturation on the elytra. *H. planus* lives in temporary grassy ponds but, because it flies so freely, it is often found singly in other habitats. In Scotland this species reaches Caithness and Sutherland, its distribution there being largely in eastern and southern lowlands, but with scattered records along the west coast, including Ailsa Craig, Arran, Bute, Colonsay, Cumbrae, Eigg, Fladda, Harris, Islay, Jura, Lady Isle, Lismore, Little Cumbrae, Rum, Skye, South Uist, and Tiree. The English distribution covers almost all areas, the exceptions being the highest parts of the Pennines and poorly recorded calcareous areas with few wetlands. The Welsh distribution is largely coastal whereas the Irish distribution is more central, both avoiding the main mountain areas. It is known from Alderney, Guernsey and Jersey. Recorded in all months, peaking in June and October.

22. *Hydroporus pubescens* (Gyllenhal) Plate 123

Length 3.2-3.8 mm. This is the only small diving beetle with the last abdominal sternite shining, being devoid of microreticulation, although the beetle's hairiness may partly obscure this character. *H. pubescens* is found in all kinds of temporary water, often also in permanent acid waters. Common across much of Britain and Ireland, particularly so in western areas where it reaches all mountain tops and the most remote and small islands, including the whole of the Scillies and St. Kilda complexes, also Alderney, Guernsey and Jersey. Gaps in its distribution are in drier parts of England and under-recorded parts of Wales. Recorded in all months, peaking in June.

23. *Hydroporus rufifrons* (Müller) Plate 124

Length 4.2-5.3 mm. The largest British *Hydroporus*, likely to be confused with *H. erythrocephalus* but easily confirmed by examination of the distinctive aedeagus. It is found in well vegetated but temporary water, usually acid or circum-neutral in pH, often in fens in river oxbows and along lake shores but also in isolated pools in hill land. This species formerly ranged across much of eastern England from Berkshire to North Northumberland. It was until recently known in Scotland in the Central Belt and Highlands to Easter Ross, and from Wales in Caerfyrddyn and Ceredigion. Natural populations now appear to be confined to the south of the Lake District and to a few sites in Kirkcudbrightshire. Recorded in all months, peaking in April and September.

24. *Hydroporus scalesianus* Stephens Plate 125

Length 1.9-2.2 mm. Easily recognised by its size, shape and reddish colour, *H. scalesianus* is associated with relict fen in palsa complexes, kettleholes and similar ancient sites, mostly in shallowly flooded moss. Fens managed for reed-cutting provide ideal conditions for it. The distribution is most unusual, ranging in Britain from a small fen in Dorset to an esker bog in Angus, the only known Scottish locality. It appears to have died out in other southern areas and across northern England, leaving a residue of coastal sites in Cumberland, County Durham, and South-east Yorkshire, and it is still frequent in the Brecks and Broads. The only Welsh site is the complex of lakes at Valley, Anglesey. These records profoundly contrast with the situation in Ireland: here it was discovered in West Meath in 1986 and has since been found in a further twenty recording vice-counties mainly in the centre but running from Limerick to Down, where it is common in interdrumlin fens. Recorded in all months, peaking in May and September.

25. *Hydroporus striola* (Gyllenhal) Plate 126

Length 3.0-3.4 mm. The most nondescript *Hydroporus*, likely to be confused with black *H. palustris* if size is not taken into account. The elytral markings are variable, typically a faint stripe running from the shoulder to the tip of the elytra, usually visible if an elytron is lifted. The habitat is marshland that may dry out in summer, dominated by rushes and with plenty of wet litter. The distribution is patchy in lowland in wetter areas in England, reaching only to the Exminster marshes in the south-west beyond the Somerset Levels. The Welsh distribution is also lowland both on the coast and inland of the mountains in Denbighshire, Flintshire, Montgomeryshire and Radnorshire. In Scotland the species ranges to the Orkneys and is common in the south but rare in the Highlands, which it skirts in the west on Islay, Lismore, Skye, and Tiree. *H. striola* is frequent in Ireland, mainly in the centre, and on the Isle of Man and Jersey. Recorded in all months, peaking in April and September.

26. *Hydroporus tessellatus* (Drapiez) Plate 127

Length 3.4-3.8 mm. When brightly marked with pale grey speckles on the elytra of this species are distinctive but its patterning can be considerably reduced, when the less heavily impressed microreticulation should be enough to distinguish it from *H. planus* and *H. pubescens*. *H. tessellatus* is found in base-rich sites ranging from saltpans at the landward edge of saltmarshes to slow-running streams on clay, and it is probably at its most frequent in farmland ponds and ditches. In England the distribution is lowland narrowing down to the coastal strip in the north-west. It is also mainly coastal in Wales, being common across

Anglesey. In Scotland *H. tessellatus* is found only along the western coast, becoming progressivly more limited to saltmarshes as it reaches Harris and Lewis. This species is common across Ireland, and is known from the Isle of Man, Alderney, Guernsey and Jersey, also on Lundy but not on the Scillies. Recorded in all months, peaking in June and September.

27. *Hydroporus tristis* (Paykull) Plate 128

Length 2.8-3.3 mm. In its most brightly coloured form, the black pronotum contrasting strongly with the red head and the elytra also red save for a dark sutural mark, this is an easily recognised diving beetle, but darker forms can be confused with other small *Hydroporus*. The habitat is generally permanent acid water amongst *Sphagnum* in peat bogs, but this species also occurs in groundwater-fed fens amongst other mosses. *H. tristis* occurs in most parts of Scotland to the northern extremity of the Shetlands and in most Hebridean islands. The rest of the British distribution clearly divides into a northern block from Glamorgan to the North-east Yorkshire coast, and a more scattered southern block, from Cornwall and Lundy to the Norfolk Broads, south of which the species is frequent only on the heaths of Dorset, the New Forest and Surrey. In Ireland *H. tristis* is frequent in most parts of Ireland but scarce in the south and east below Down. *H. tristis* is known from the Isle of Man. Recorded in all months, peaking in June.

28. *Hydroporus umbrosus* (Gyllenhal) Plate 129

Length 2.5-2.8 mm. This small and dark species could be confused with *H. striola* if its size is not taken into account. Its habitat is in permanent water, usually in mesotrophic or base-rich fen conditions in marshes, ponds and ditches over peat. *H. umbrosus* is common on low ground in southern Scotland and along the east coast to Aberdeenshire, also on Speyside, but it is scattered elsewhere, ranging to the Orkneys. It is also known from most of the Hebrides, and from four of the Clyde Isles including Holy Island but not Arran. A major block of records runs north of a line from Glamorgan to the Yorkshire coast in lowlands. In common with its absence from the Channel Isles this species is absent from south-west England and around the Severn Estuary, except for Priddy, Shapwick Heath, and Westhay in North Somerset. Other southern records are clustered over heathland and peaty areas, rare in Berkshire, and from Dorset, the New Forest and the Surrey heaths, Wicken Fen, the Brecks, Greensand and the Broads of Norfolk, with older records from the Central Weald. Records in Ireland are also scattered with a few clusters, in the Burren, the northern karst around Lough Mask and in the interdrumlin mires of Armagh and Down. *H. umbrosus* is frequent on the Isle of Man. Recorded in all months, peaking in May.

17. *NEBRIOPORUS* Régimbart

Several genera of small diving beetles have species with a pale body colour and the elytra with dark stripes; those with small teeth at the rear of the elytra belong to *Nebrioporus*. They are associated with exposed substrata in ponds, lakes and rivers. *Nebrioporus depressus* and *N. elegans* can only safely be determined by examination of the male aedeagus – body shape, fore claws and the width of the fore and mid tarsi differ between males and females of these taxa. The other two species need not be sexed or dissected. *N. canaliculatus* was until recently placed in a subgenus, *Zimmermannius* Guignot, which Toledo (2009) has reduced to a species-group: the large size and pale colour of *canaliculatus* will command attention even if the three weak ridges on each elytron, unique among British diving beetles, are not immediately visible.

Key 16. The species of *Nebrioporus*

1. Each elytron with a small tooth projecting from the margin near the tip (Fig. 240) 2

- Elytra without teeth at rear (see couplet 13 of key 6 for more information) .... 2. *Nebrioporus canaliculatus* (Lacordaire) (p. 92)

240

2 Length 4.3 mm or less; blotches on the hind margin of the pronotum extending forwards more than a third of the way across the pronotum; the elytral stripe next to that of the suture usually interrupted behind the middle, the second stripe being continuous, but some specimens may be much darker overall (Fig. 241); sides of pronotum broadest at the rear, meeting elytra at a slight angle in dorsal view ....
.... 1. *Nebrioporus assimilis* (Paykull) (p. 92)

241

- Length 4.5 mm or more; blotches on the hind margin of the pronotum extending less than a third of the way across the pronotum; first elytral stripe continuous but second with a distinct gap midway along, and merging in part with the third stripe to form a distinctive pale spot visible even in darker individuals (Fig. 242); sides of pronotum strongly curving in at rear in females (Fig. 243), meeting elytra at a distinct angle, male pronotum rear not so strongly curved (Fig. 244) but still more than in *N. assimilis* 3

242

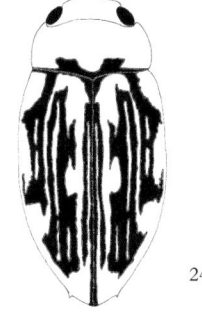

243   244

3. Tip of the aedeagus wide (Fig. 245), though with variants that are narrow; inner fore claw of male strongly curved in outer third, i.e. "hooked" (Fig. 246) ....
... 3. *Nebrioporus depressus* (Fabricius) and the intermediates (p. 92)

- Tip of the aedeagus always narrow (Fig. 247); male fore claws evenly curved (Fig. 248) .... 4. *Nebrioporus elegans* (Panzer) (p. 93)

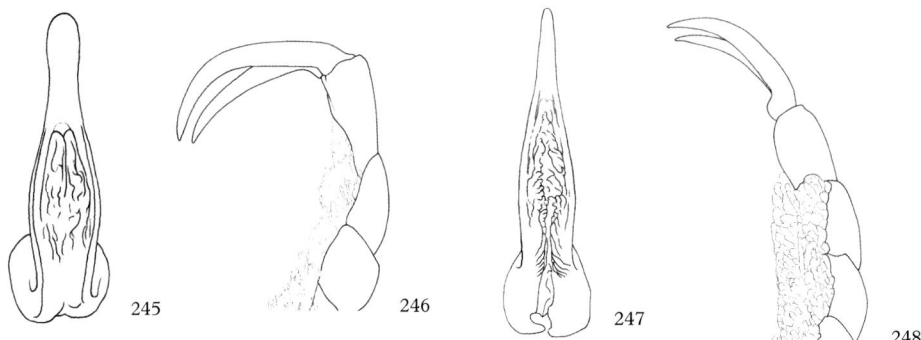

1. *Nebrioporus assimilis* (Paykull) Plate 130

Length 4.0-4.3 mm. *N. assimilis* is confined to permanent, still water and is typical of lake shores with fixed gravel beds covered with shore weed (*Littorella uniflora* L.). It can also occur in man-made habitats such as canals, reservoirs and disused workings. It is a lowland species reaching south to North Essex, being wholly absent from the south of Wales and England south of the Severn. In the north it reaches the Shetlands and is common on the Hebrides where it may be almost entirely black. The Irish distribution is also mainly northern and central, but it does reach North Kerry. There is one old record for the Isle of Man. Recorded in all months, peaking in June and August.

2. *Nebrioporus canaliculatus* (Lacordaire) Plate 131

Length 4.8-5.8 mm. *N. canaliculatus* is a mobile species found in shallow water with little or no vegetation in gravel pits. It was first reported in England at Dungeness in East Kent in 1999 since when it has been found in other parts of the same area. Recorded in May and from August to October.

3. *Nebrioporus depressus* (Fabricius) and intermediates with *N. elegans* Plate 132

Length 4.5-5.2 mm. The *N. depressus-elegans* complex appears to be based on two subspecies that can interbreed to produce stable intermediates in Britain. The true *N. depressus* has a spatulate aedeagus but intermediates with narrow aedeagi resembling *N. elegans* can be found, particularly in southern Scotland. The habitat is typically vegetation-free deep water in larger lakes in the south of its range but *N. depressus* also occurs in northern Scottish rivers in the absence of *N. elegans*. *N. depressus* has been recorded from the Orkneys, Mull and Islay. In England *N. depressus* is confined to the Lake District, the population in the neighbouring Talkin Tarn having probably become extinct; it is still abundant in Lake Windermere, and is also known from Blea Tarn, Grasmere and Rydal Water. An intermediate form is known from Tal-y-Llyn, Meirionydd. *N. depressus* is common across Ireland in loughs, rivers and quarry ponds. Neither it nor *N. elegans* occurs on the Isle of Man. Recorded in all months, peaking in July.

4. *Nebrioporus elegans* (Panzer) Plate 133

Length 4.4-5.0 mm. *N. elegans* is typical of lowland rivers, and also of man-made lakes, always over an exposed substratum, usually with thin vegetation – or beds of water-crowfoot (*Ranunculus aquatilis* L.) when in rivers. It is common across most of Britain reaching the Orkneys but it is rare in the Highlands. Other island records are for Bute, Islay, Mull, Rum, and Skye, also the Isle of Man, Anglesey, the Isle of Wight and Jersey. In Ireland recent finds in Wicklow and North Tipperary indicate an invasion by *N. elegans*. Recorded in all months, peaking in August.

18. *OREODYTES* Seidlitz

These are species with a pale upper side, 2.9-5.0 mm long. The elytra are usually striped though the stripes may coalesce into blotches. The pronotum has lateral depressions similar to those of *Graptodytes* and *Stictonectes*, however, these can vary – even in the same species – from being shallow with the same microreticulation as the rest of the pronotum to having the trough of each furrow shining without microreticulation and sharply defined by ridges. The pronotal spotting is particularly variable. All three running water species can be found together in many northern rivers. With the exception of the elytral modification of female *O. alpinus* sexual differences are slight. Examination of the genitalia might only be necessary to differentiate the males of *O. alpinus* and *O. davisii*: the tip of the aedeagus is easily damaged by dissection and the genitalia are best extruded from freshly killed beetles.

Key 17. The species of *Oreodytes*

1. Length exceeding 3.7 mm 2

- Length 3.6 mm or less 3

2. Elytral stripes reach forwards to leave only a narrow yellow margin (Fig. 249); female with the tips of the elytra flared (Fig. 249); aedeagus slightly constricted immediately behind the tip to produce a small process (Fig. 250) ....
   .... 1. *Oreodytes alpinus* (Paykull) (p. 94)

- All bar the third of the elytral stripes falling short to leave a wide yellow bar at the front of the elytra (Fig. 251); rear of elytra without expansion in either sex; aedeagus simple at the tip (Fig. 252)
   .... 2. *Oreodytes davisii* (Curtis) (p. 94)

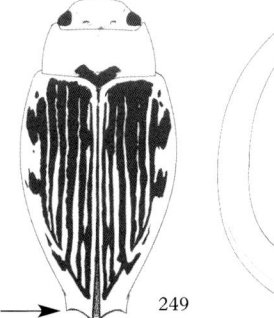
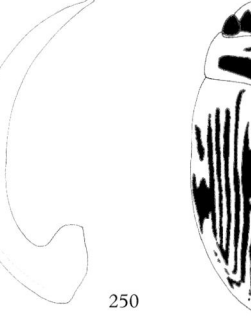
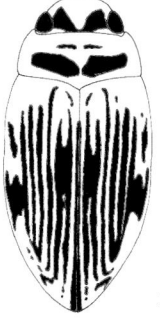
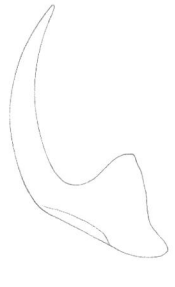

249 250 251 252

3. Body shape long oval (Fig. 253); abdomen always black below; trochanters of the hind legs with a few small punctures and sparse short hairs (Fig. 254); elytra with large punctures scattered thickly between those forming furrows .... 4. *Oreodytes septentrionalis* (Gyllenhal) (p. 95)

- Body shape round oval (Fig. 255); abdomen orange or black below; trochanters of the hind legs with many coarse punctures supporting long hairs (Fig. 256); hardly any punctures on the elytra other than those forming furrows .... 3. *Oreodytes sanmarkii* (Sahlberg) (p. 95)

This species has the greatest tendency for the elytral stripes to coalesce into blotches, as shown in Fig 255, which should be compared with the striped form in Plate 136.

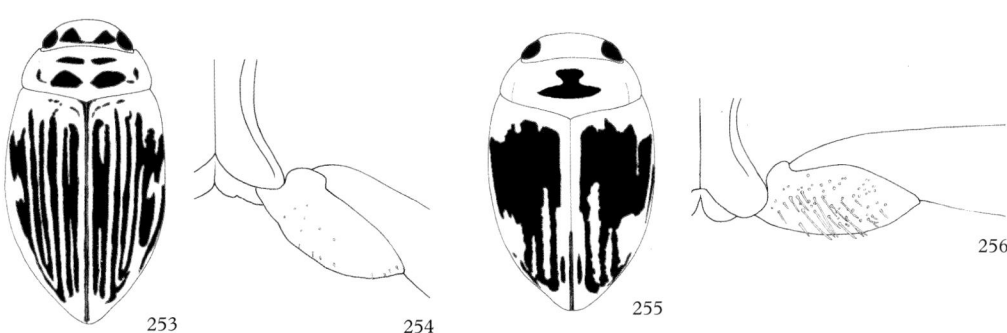

253   254   255   256

1. *Oreodytes alpinus* (Paykull) Plate 134

Length 4.2-5.0 mm. This is the largest Scottish *Oreodytes*, easily recognised because of its special habitat, which is never occupied by the similar *O. davisii*. The expansion of the female elytra is also distinctive. Although the first specimen was detected in Scotland in 1985 in a river, *O. alpinus* is associated with the ends of large lochs subject to intense wave action resulting in unstable sand and swirling debris devoid of vegetation. Known in Britain only from Caithness and East Sutherland. Recorded in March and from May to September, peaking in August.

2. *Oreodytes davisii* (Curtis) Plate 135

Length 3.8-4.5 mm. This species is a little smaller than *O. alpinus*. The intensity of the dark markings varies considerably – typically the elytral stripes fall short of the front edge leaving a wide transverse yellow bar that is distinctive in the field. *O. davisii* is confined to running water, most typically associated with side pools around rock sills, but also found on unstable shingle. The main part of the English distribution runs north from Derbyshire along the Pennines, being almost absent in the Lake District but descending to sea level from North-east Yorkshire northwards on the east coast, with records in the west from streams in Herefordshire. In Wales the distribution is in fast rivers in Breconshire, Caerfyrddyn, Caernarfon, Ceredigion, Denbighshire, Meirionydd, Monmouthshire, Montgomeryshire, Pembrokeshire, and Radnorshire, with old records for Glamorgan and Pembrokeshire. In Scotland the species is frequent in the Southern Uplands but more lowland and scattered further north to Caithness, being known from Rum and Skye. In Ireland found in mountain streams in the north and east in Antrim, Down, County Dublin, Londonderry, Tyrone, and Wicklow. Recorded throughout the year, peaking in July.

3. *Oreodytes sanmarkii* (**Sahlberg**) Plate 136

Length 2.9-3.3 mm. It should rarely be necessary to invoke the character of the hairy metatrochanters that set this small species apart from other *Oreodytes*. This is a common species in shallow streams, usually on gravel beds, but its distribution is restricted, being absent or very rare over much of lowland England in the south and the east. There is an isolated cluster of records north of London in Buckinghamshire, south Cambridgeshire, and Hertfordshire. *O. sanmarkii* is common across Wales including Anglesey. In Scotland it is also common, reaching the Orkneys, and is known from Arran, Bute, Eigg, Islay, Jura, Mull, Raasay, Rum, and Skye, but is absent from the Outer Hebrides. Nor is it found on the Isle of Man. The Irish distribution is mainly associated with coastal mountain systems, but it is common over most of Northern Ireland. Recorded throughout the year, peaking in August.

4. *Oreodytes septentrionalis* (**Gyllenhal**) Plate 137

Length 3.2-3.6 mm. This species could be confused with *O. sanmarkii*, but is slightly larger, always black underneath and always with sharply defined elytral stripes. *O. septentrionalis* is found in fast rivers over unstable gravel beds, and occurs in large, steep-edged lakes. The British distribution divides into three distinct sections, the south-west of England, Wales and the Welsh Marches, and Scotland and England from Ilkley Moor in Mid-west Yorkshire northwards, with an outlying record, in 1984, for the River Lymn in North Lincolnshire. *O. septentrionalis* occurs on Islay, Mull and the Isle of Man. The Irish distribution is very similar to that of *O. sanmarkii*, slightly less frequent overall but more common in the south-west. Recorded in all months except December and February, peaking in July.

19. *PORHYDRUS* Guignot

This small Palaearctic genus reaches Britain and Ireland with one fairly distinctive small and striped species. The lobed metacoxal processes depicted in Key 6 are typical of the genus.

1. *Porhydrus lineatus* (**Fabricius**) Plate 138

Length 3.0-3.5 mm. If the elytral stripes are faint, this beetle could be confused with *Hydroporus obscurus*, being reddish and having a similar tear drop body shape, broad at the front, widest before the middle and tapering to the rear. Its lobed metacoxal processes (see Key 6) will distinguish it from *Hydroporus* species. This species is particularly typical of grazing fen ditches and other permanent and patchily vegetated, base-rich waters. In England and south Wales the distribution reflects this association with coastal fenland, and *P. lineatus* is also frequent on the Cheshire Plain and on fens on the inner parts of the Humber, with scattered records across the intervening area. Another centre is in Dumfries and Galloway, mainly in lake fens. Outlying records stretch the distribution from Goss Moor in West Cornwall to Fithie Loch in Angus. *P. lineatus* occurs on Anglesey, the Isle of Man, Guernsey and Jersey. It is common in central Ireland but it is rare in the south and absent from much of the north. Recorded throughout the year, peaking in May and September.

20. *SCARODYTES* des Gozis

One of the features claimed to distinguish this genus is the presence of lateral furrows on the pronotum but they are so weak that it is expected that most specimens will ultimately run to couplet 28 in Key 6 rather than to couplet 20, hence its appearing in the key twice. The broad oval body shape and the strong dorsal patterning on a pale background contrasting with a dark underside should distinguish members of the genus, which is also characterised by the upper surface's fine hair and puncturation, each puncture being no more than a puncture's diameter from the next.

1. *Scarodytes halensis* (Fabricius) Plate 139

Length 3.8-4.3 mm. The first and second elytral stripes either side of the sutural stripe are usually joined together at three points to give ladder-like features to the pattern. Males are distinguished from females by slightly longer fore claws and wider fore and mid tarsi, but a useful character is provided by the sexual differences in the colouring of the underside, females being black except for the reddish brown abdominal sternites whereas males are completely black save for the edges of the sternites. These characters together should serve to set this species apart from other pale diving beetles with elytral stripes. The habitat is also specific:- base-rich, even polluted, pools and slow running ditches with sparse vegetation, including recently cleared dykes and trunk road balancing lagoons. Furthermore the distribution is unusual in being compact and defined within an area bounded by a line from Yorkshire on the coast at Skipsea Brough to Castleford, south-west to Croome Park in Worcestershire then eastward through Oxfordshire, north of London, through East Suffolk and to the Norfolk coast at Wells-next-the-Sea. There are old records from the Norfolk Broads to East Kent. Recorded from January to November, peaking in April and August.

21. *STICTONECTES* Brinck

Another small Palaearctic genus, unmistakeable in that most species are black with transverse wavy lines on the elytra. The individual species are difficult to identify but it seems that only *S. lepidus* has reached Britain and Ireland.

1. *Stictonectes lepidus* (Olivier) Plate 140

Length 3.1-3.4 mm. *S. lepidus* is typically found on hard dark surfaces, either peat or rock, in still water and in streams with stepped pools, but it can also occur over gravel beds. It was once widespread north to the Orkneys but there are now many areas, especially on the east sides of both Britain and Ireland, where the species can no longer be found. Strongholds are small streams in the west, usually running directly to the sea without connecting to major river systems, the Pennines on the Millstone Grit, and peatworking areas. It has recently been recorded on the Isle of Man, Anglesey, Guernsey, Jersey, Clare and from Arran, Coll, Cumbrae and Islay. There is an old record for Herm. Recorded from January to November, peaking in June and August.

22. *STICTOTARSUS* Zimmermann

1. *Stictotarsus duodecimpustulatus* (Fabricius) Plate 141

Length 5.2-5.7 mm. A spotted diving beetle quite unlike any other in Britain and Ireland. It is typical of rivers, canals and rocky shores in lakes and reservoirs, generally with vegetation sparse or absent. Widely distributed in Ireland and Britain, reaching all of the Outer Hebrides and Caithness but not the Orkneys. Also known from the Isle of Man and Jersey. Recorded in all months, peaking in August.

23. *BOREONECTES* Angus

1. *Boreonectes multilineatus* (Falkenström) Plate 142

Length 4.0-4.8 mm. A dark, striped species probably best signalled by its habitat, shallow montane lakes with a mixture of peat and gravel as the substratum. It occasionally occurs on low ground in the Southern Uplands and from North Aberdeenshire northwards. In England it is known only from Whernside and Fountains Fell in Mid-west Yorkshire. In Wales it is confined to Snowdonia in Caernarfon and Meirionydd. It occurs in Scotland from the Southern Uplands to the Shetlands, where it is common, and it is known on the main Hebridean islands and Arran. In Ireland *S. multilineatus* is known from Antrim, Down, Fermanagh, West Galway, South Kerry, Londonderry, West Mayo, Monaghan, Sligo, Waterford, and Wicklow, though there are no modern records for mountains south of Galway. Recorded from March to November, peaking in June and September.

24. *SUPHRODYTES* des Gozis

Suphrodytes is easily recognised as a large hydroporine until now associated with considerable variation in the extent of its elytral pattern (see Figs 257-260). Recent studies of the DNA of Dytiscidae have established that there are two species previously represented by *dorsalis* sensu lato. Both species occur in lowland pools and fenland ditches in part shade, the two often coexisting. The complex is widespread in lowland England ranging to the extreme north at Kingmoor, Carlisle and, as 19th Century records, to North Northumberland, still being frequent in the south of that county, but has never been reported from Scotland. In Wales it is confined to the eastern edge in Denbighshire and Monmouthshire, with an outlying 19th Century record in Glamorgan. The Irish distribution is compact, running in a broad band from Limerick to Belfast. The complex has not been reported from any island. Adults of this complex have been found throughout the year, peaking in May and August.

In addition to wider segments on the fore and middle tarsi, the males are distinguished from females by a tooth on the inner fore claw. The best way to distinguish the two species is to start with the more strongly marked specimens in any series, assigning them to one or other of the species and using their size and shape as a template against which to judge the rest of the series. Given that these species can coexist, dissection of dark males may still be advisable, the aedeagi being most usefully viewed dorsally.

The DNA analysis referred to above (Johannes Bergsten and Anders Nilsson, pers. comm.) also established that *Suphrodytes* is so strongly linked to *Hydroporus* that it may not justify generic status.

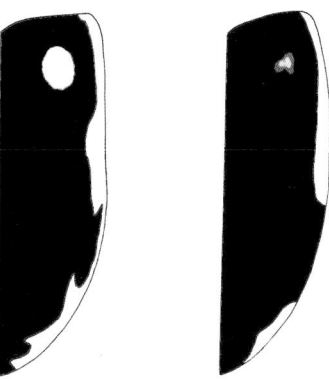

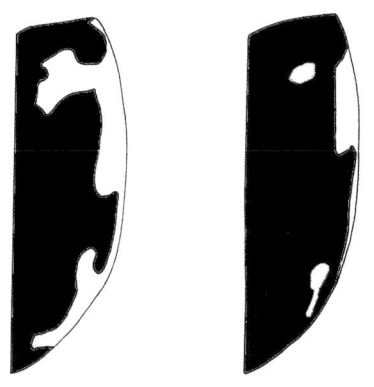

Figures 257 and 258 Variation in the elytral pattern of
Suphrodytes dorsalis

Figure 259 and 260 Variation in the elytral pattern of
Suphrodytes figuratus

Key 18. The species of *Suphrodytes*

1. Elytra, as measured along the suture, exceeding 3.5 mm in length; head typically entirely red with dark patches beside eyes greatly reduced; pale patterning of elytra at most comprising a longitudinally extended oval spot near the shoulder and pale edges; aedeagus relatively broad with a more robust base (Figs 261 and 262) .... 1. *Suphrodytes dorsalis* (Fabricius)

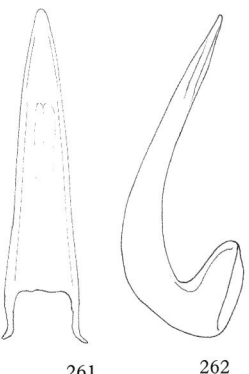

261   262

- Elytra 3.5 mm long or less; head usually with large dark wedge-shaped marks beside eyes; elytral pale pattern at its greatest development comprising two large crescents connected by a wide lateral band, but often reduced to a pair of irregularly shaped spots near the shoulder, sometimes wholly dark; aedeagus narrower than in *dorsalis*. with a less robust base (Figs 263 and 264) ....
.... 2. *Suphrodytes figuratus* (Gyllenhal)

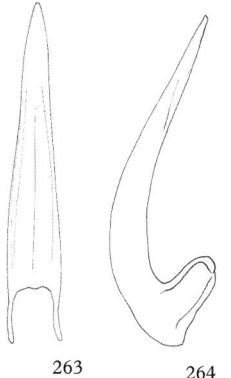

263   264

1. *Suphrodytes dorsalis* (Fabricius) Plate 143

Length 4.6-5.3 mm. This species has so far been recognised from: England – Dorset, East Norfolk, Cambridgeshire, Worcestershire, North Lincolnshire, South-east and Mid-west Yorkshire, Cumberland; Ireland – Limerick, Laois, South-east Galway, Meath, East Mayo, Armagh, Down.

2. *Suphrodytes figuratus* (Gyllenhal) Plate 144

Length 4.3-4.6 mm. This species has so far been recognised from: England – North Somerset, Dorset, East and West Sussex, West Kent, Surrey, North Essex, West Suffolk, East and West Norfolk, Cambridgeshire, Herefordshire, North Lincolnshire, Leicestershire, Mid-west Yorkshire, South Lancashire, County Durham, South Northumberland; Wales – Monmouthshire; Ireland – Limerick, South-east Galway, Meath.

25. *HYDROVATUS* Motschulsky

These are small, rounded and flat insects, the only diving beetles in Britain and Ireland to have the tips of the elytra drawn out into a point. Other important characters are the blunt prosternal process, with a strong step in its neck, the incisions into the rear edges of the metacoxal processes and the short wide antennal segments. Dissection of males is advised for those specimens that have obscure spots on the elytra. Males can be recognised by the presence of a stridulatory file on the outer half of the front edge of each metacoxal plate. The aedeagi are most easily identified in side view. The female vulval sclerites are drawn out into an ovipositor that looks the same in both species.

Key 19. The species of *Hydrovatus*

1. Elytra either without markings or vaguely mottled; male with front of head, viewed from above, angular (Fig. 265), female head shape an arc; first and second abdominal sternites with a series of small grooves, without large punctures (Fig. 268); aedeagus with the tip straight (Fig. 270) .... 1. *Hydrovatus clypealis* Sharp (p. 100)

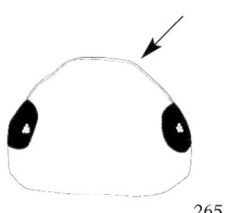

265

It is unnecessary to remove the hind leg as in the figure, but it must be moved to get a good view of at least the first sternite; the tip of the aedeagus is very thin and easily broken.

- Elytra with two large spots with extensions to the outer edges (Fig. 267), though sometimes poorly defined; front of head, viewed from above, an arc in both sexes (Fig. 266); first and second abdominal sternites with large punctures (Fig. 269); aedeagus with the tip hooked (Fig. 271) ....
.... 2. *Hydrovatus cuspidatus* (Kunze) (P. 100)

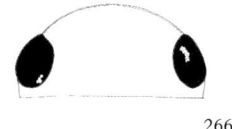

266

267

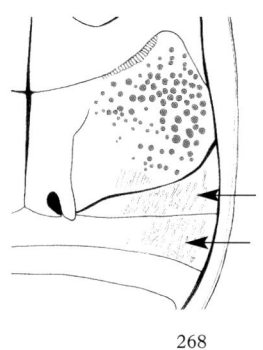

268

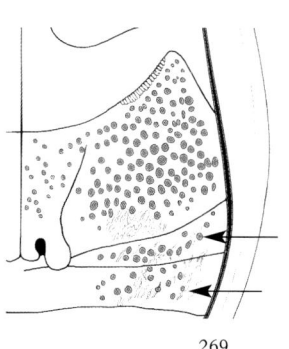

269

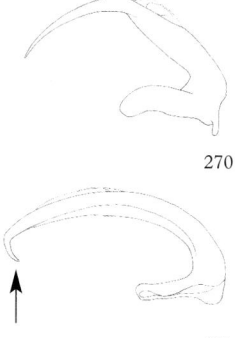

270

271

1. *Hydrovatus clypealis* Sharp Plate 145

Length 2.2-2.5 mm. Apart from the array of characters given to distinguish this insect from *H. cuspidatus* the habitat can also be characteristic, ranging from the edges of ponds on soft mud to floating rafts of vegetation such as Marsh St. John's-wort (*Hypericum elodes* L.). In England this species is confined to the south along the coast from North and South Somerset, out to St. Mary's in the Scillies and eastwards to Dungeness, the Thames Marshes and recently further north into South Essex and East Suffolk. Despite this *H. clypealis* is rarely found in brackish water. There are old records for Wales in Glamorgan, Pembrokeshire and Caerfyrddyn, with a record for Anglesey in 2002. In Ireland the first record, from Waterford, also in 2002, with further records from South Tipperary and Wexford. Additional island records are for Guernsey and Jersey, and for Brownsea in Dorset. Records are from March to December, peaking in May and September.

2. *Hydrovatus cuspidatus* (Kunze) Plate 146

Length 2.4-2.7 mm. The males are much more shining than the females. This species is associated with man-made habitats, usually in the extreme edge of thinly vegetated ditches and pools over exposed clay. Records so far are from drainage ditches on levels in East Kent, East Suffolk and East Norfolk. Recorded in April, May, June, August, and September.

26. *HYGROTUS* Stephens

This genus has nine species in Britain, seven in Ireland. They are small, 2.2-5.5 mm long, all but the smallest having the elytra striped. All are deep-bodied and either globular or elongate in outline. The division into two subgenera is based mainly on the presence of a raised front margin to the head on the clypeus in the subgenus *Hygrotus*, but the major difference in body shape cuts across this division, both subgenera having short and globular species. The more globular species without the clypeal ridge are *H. confluens* and *H. nigrolineatus*: they are easily distinguished from the rest by large parts of their elytra being transparent, allowing one to see the wings and two large tracheae that are tracked by the second and fourth elytral stripes – the pale areas on other species are opaque. The puncturation of the elytra and the pronotum provides an important character for separating two species-pairs. However the dimorphism of the females of *H. quinquelineatus* and *H. impressopunctatus*, with strongly microreticulate matt forms lacking larger punctures, has led to misidentifications in the past. Simple variations in the patterning of the thorax and elytra provide characters that are just as reliable as structural ones, and are more easily described. It should not be necessary to dissect the genitalia in order to identify any species: anyone doing so is warned that the aedeagi have long thin points that are easily broken off.

Key 20. The species of *Hygrotus*

1. Body globular; length 2.0-4.0 mm 2

\-   Body elongate; length size 4.0-5.0 mm 7

2. Abdomen black, strongly contrasting with partly transparent elytra with the pale yellow epipleurs; front of head (clypeus) without a raised rim 3

- Underside uniformly red or yellow, upper surface's ground colour an opaque yellow; front of head with a raised rim 4

3. Counting from the suture, at least the second or fourth black lines, or both, starting just behind the front edge (Fig. 272); length 3.6 mm or more; male inner fore claw with a straight section and longer than outer claw (Fig. 273) .... 7. *Hygrotus nigrolineatus* (von Steven) (p. 105)

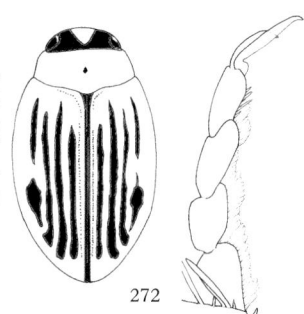

- Black lines beginning a third of the way down the elytra, leaving the shoulders entirely pale (Fig. 274); length 3.6 mm or less; male inner fore claw about the same length as the outer one (Fig. 275) .... 5. *Hygrotus confluens* (Fabricius) (p. 104)

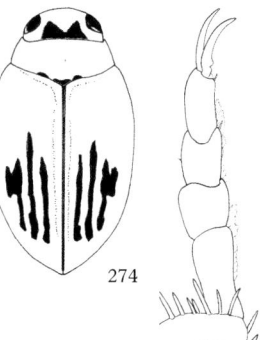

4. Length at least 4.0 mm; elytra predominantly pale with longitudinal black lines 5

- Length 2.0-3.6 mm; elytra usually predominantly dark with irregular paler patches but sometimes predominantly pale with darker blotches or lines 6

5. Large punctures (over three times the size of the smaller ones) on the elytra sparse and far outnumbered by the others (viewed at x 30) (Fig. 276); dark stripes nearest the elytral suture not reaching the front margin and usually broken into two bars (Fig. 277) .... 4. *Hygrotus versicolor* (Schaller) (p. 104)

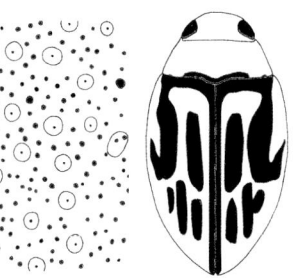

- Large punctures on the elytra as numerous as small ones (Fig. 278); dark lines nearest the suture of the elytra reaching very near to the front margin (Fig. 279) .... 3. *Hygrotus quinquelineatus* (Zetterstedt) (p. 104)

Females have the same puncturation irrespective of whether they are of the microreticulate form or of the shining, male-like, form.

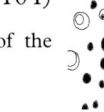
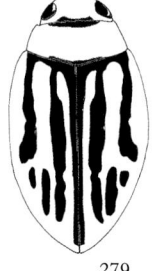

6. No pale band across pronotum; elytra chestnut with paler patches near the anterior margin and the tip (Fig. 280); not more than 2.6 mm long .... 1. *Hygrotus decoratus* (Gyllenhal) (p. 103)

Graptodytes pictus and *Hydrovatus cuspidatus* have a similar elytral pattern to this species.

280

- Pale band across pronotum; elytra mainly black with irregular dark pattern near the anterior and lateral margins, but sometimes with the black marks broken up into bars (Fig. 281); not less than 2.7 mm long .... 2. *Hygrotus inaequalis* (Fabricius) (p. 103)

281

7. Males and shining females with strongly punctured striae on the elytra and much of the upper side covered with large coarse punctures interspersed with about equal numbers of small punctures; pronotum with a dark mark spreading out from about halfway down the central incision, usually connecting to a dark band along the rear margin (Figs 282 and 283); male fore claws equal in length, the inner one more strongly curved than the outer (Fig. 284) .... 6. *Hygrotus impressopunctatus* (Schaller) (p. 105)

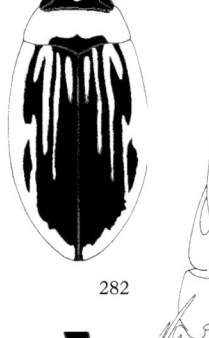

282

284

283

- All specimens with upper sides covered with very fine punctures irrespective of whether the beetle is matt or not; pronotum with a round or diamond-shaped dark spot around the central incision, if connected at all to the rear edge then only by a narrow central streak; male inner fore claws more curved than the outer ones ...
.... 8

If matt without striae check both sides of this couplet for the pronotal mark, as *H. impressopunctatus* has a matt female form that resembles *H. parallellogrammus* in puncturation.

8. Length 4.5 mm or longer; counting out from the suture the second elytral stripe reaches the front margin within a dark bar (Fig. 285); male inner fore claw broad and pointed (Fig. 286); females either matt or male-like ....
.... 9. *Hygrotus parallellogrammus* (Ahrens) (p. 105)

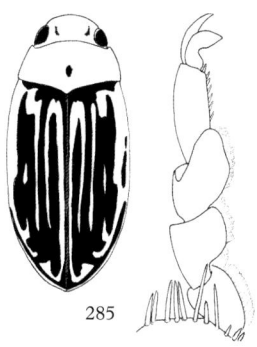

285

286

- Length 4.0 mm or less; elytral stripes stopping short of the front margin (Fig. 287); male inner fore claw blunt, looking as if the tip may have broken off (Fig. 288); all females matt ....
.... 8. *Hygrotus novemlineatus* (Stephens) (p. 105)

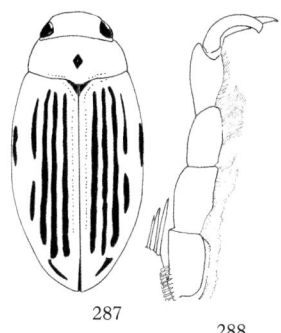

287

288

Subgenus *Hygrotus* Stephens

1. *Hygrotus decoratus* (Gyllenhal) Plate 147

Length 2.2-2.6 mm. Once this species has been determined as a *Hygrotus* there should be no question as to its identity. However its size, shape and elytral pattern resemble those of *Graptodytes pictus*, and *Hydrovatus cuspidatus*, although wider and flatter, is similarly marked. The *Graptodytes* have pronotal slots and the *Hydrovatus* has the elytra drawn out into a blunt process at the tip, rather than being rounded. *H. decoratus* occurs in richly vegetated and permanent ponds and ditches, and in mossy swamps. It is frequent in south-east England, East Anglia and Cheshire but restricted elsewhere, rare in the New Forest, Herefordshire, Shropshire and Leicestershire. There is a northern population centred on the Thorne Waste and Askham Bog, and ranging north to Catterick in North-west Yorkshire. In Wales there are old records for Caernarfon and Glamorgan, and modern ones for Denbighshire. In Ireland *H. decoratus* is known only from two adjacent lakes in Limerick. Recorded in all months, peaking in April and September.

2. *Hygrotus inaequalis* (Fabricius) Plate 148

Length 2.7-3.5 mm. *H. inaequalis* is recognisable as the only small globular and dark diving beetle in Britain and Ireland. It can be very dark but the white bar across the pronotum is usually visible in life. *H. inaequalis* occurs in a wide range of permanent water habitats, but often in very shallow water. It occurs over much of Britain and Ireland, being rare in hill land and absent from much of the Welsh mountains and the Scottish Highlands. Similarly it is mainly found inland in Ireland, being absent from much of the coastal mountain systems. *H. inaequalis* occurs on most islands with permanent water from the Channel Isles north to the Orkneys. Recorded in all months, peaking in August.

3. *Hygrotus quinquelineatus* (**Zetterstedt**) Plate 149

Length 3.1-3.6 mm. Unlike *H. versicolor* this species often has a mixture of females, some dull and microreticulate and others with the same appearance as the males. *H. quinquelineatus* is found in lowland unproductive lakes including the phosphate-deficient marl loughs of the Burren. It is found along deep parts of the shoreline among thin vegetation, often under overhanging trees. There are records for rivers, canals and reservoirs but this seemingly flightless species is more typical of natural still waters. Modern records in England are from Cheshire, Cumberland, Derbyshire, North and South Lincolnshire, Nottinghamshire, and South-west Yorkshire, but old records indicate a much wider distribution in the past. Scottish records are mainly for the south-west and for east of the Highlands. *H. quinquelineatus* is common across much of Ireland. There are no records for any islands other than the British and Irish mainlands. Recorded in all months, peaking in June and August.

4. *Hygrotus versicolor* (**Schaller**) Plate 150

Length 3.1-3.6 mm. The elytral pattern and the large punctures much sparser relative to the fine punctures should clearly set this species apart from *H. quinquelineatus*. *H. versicolor* does not appear to have dimorphic females. Microreticulate females with few large punctures should be checked for *quinquelineatus* using the elytral pattern differences described in key 20. The habitat of *H. versicolor* is amongst thin vegetation in ponds, canals and drainage ditches, usually on an exposed substratum of clay or peat. In England it is common in the lowlands to the southern edge of the Lake District but it is absent from the south-west beyond the Exminster marshes in South Devon. *H. versicolor* is scarce in Wales with records from Anglesey, Caerfyrddyn, Caernarfon, Glamorgan, Meirionydd, Monmouthshire, and Montgomeryshire. There are small clusters of records in Kirkcudbrightshire and in West Meath. Recorded in all months except February, peaking in May and August.

Subgenus *Coelambus* Thomson

5. *Hygrotus confluens* (**Fabricius**) Plate 151

Length 3.2-3.6 mm. *H. confluens* is confined to newly created and highly disturbed ponds, rarely amongst vegetation but often in brackish water and in the polluted conditions of rubbish disposal sites, cooling waters from heavy industry, effluent treatment beds, rockpools fouled by sea birds, and farm ponds heavily fouled and churned up by livestock. Females and males of this species are both shining. *H. confluens* occurs over most of lowland England. It appears to be absent from mid Wales. In Scotland the species is rare, ranging north to Angus, with most records in the Central Belt. In addition to Anglesey, the Scilly Isle St. Mary's, the Isle of Wight, and Guernsey *H. confluens* also reaches the small offshore islands of Lady Isle near Troon in Ayrshire, the Isle of May, Fife and Inisheer, Clare. It can also colonise pools on high ground such as Thorpe Pasture and on the summit of Snake Pass in Derbyshire. *H. confluens* is scarce in Ireland, with records from Antrim, Clare, Down, Dublin, Fermanagh, Meath, Mid Cork, North Kerry, North Tipperary, South-east and West Galway, and Wexford. Recorded in all months, peaking in April and August.

6. *Hygrotus impressopunctatus* (Schaller) Plates 152 and 153

Length 4.1-4.5 mm. This is the most common of the large *Hygrotus*, typically living in rich fen in lowland lakes, ponds and ditches, but also to be found amongst *Sphagnum* in mesotrophic fen. The matt female form *lineellus* Gyllenhal is often misidentified as *parallellogrammus* because it lacks the large punctures of the male and the male-like female. *H. impressopunctatus* is frequent in much of lowland England apart from the south-west. Welsh records are more scattered and mainly coastal. In Scotland it ranges as far as the north of Fife. It is scarce across much of Ireland, being common only from the Burren in a broad band across to Wexford. Island records include Alderney, Jersey, the Isle of Man, Anglesey, the Isle of Man and the Isle of May. Recorded in all months, peaking in May and August.

7. *Hygrotus nigrolineatus* (von Steven) Plate 154

Length 3.6-3.9 mm. *H. nigrolineatus* is only likely to be mistaken for the slightly smaller and less heavily marked *H. confluens* with which it can occur in the same habitat, recently created or disturbed still water with minimal vegetation, often polluted but rarely brackish. Sites are vacated when vegetation takes over. This species is dimorphic with the matt form of the female being common; matt males also occur. *H. nigrolineatus* was found first in England in 1983 in East Kent since when it has spread to South Devon in 1996, to South Northumberland in 2004 and to central Glasgow in Lanarkshire in 2010. It is now most frequent in the English Midlands. *H. nigrolineatus* was first recorded in Wales in Flintshire in 2000. Recorded from February to November, peaking in April and September.

8. *Hygrotus novemlineatus* (Stephens) Plate 155

Length 3.5-4.0 mm. *H. novemlineatus* should not be confused with any other species so long as size is taken into account. All females are of the matt form. It is associated with permanent water in exposed lakes, usually with light-coloured, finely particulate bottom substrata, usually either sand or silt, sometimes interspersed with peat. Such sites are often base-poor but the species also occurs on calcareous lake marl in Ireland. *H. novemlineatus* has died out in southern England being known now from North-west Yorkshire, County Durham, South Northumberland and Cumberland. The Scottish distribution is patchy, mainly in lakes in the west of the Southern Uplands, then from Loch Lomond north to the Orkneys. Modern island records are for Islay and Tiree. Modern Irish records are from Meath, North Kerry, Sligo, South-east Galway, West Donegal and West Meath. Recorded from March to November, peaking in August.

9. *Hygrotus parallellogrammus* (Ahrens) Plate 156

Length 4.5-5.5 mm. This species is narrower and more brightly marked than *H. impressopunctatus*. It is confined to brackish water, but there are occasional inland records, some of which may have stemmed from misidentification of matt female *H. impressopunctatus*. Both types of female occur in this species, the most matt one being more frequent. Apart from the Severn Estuary this species is found almost continuously along the coast from the Isle of Wight and South Hampshire to the Humber in South-east Yorkshire. There are old and doubtful records from Ireland. Recorded from January and from March to November, peaking in May and September.

27. *HYPHYDRUS* Illiger

Two unmistakable species, 3.9-5.3 mm long, and globular.

Key 21. The species of *Hyphydrus*

1. Colour red, with or without mottling ....
.... 2. *Hyphydrus ovatus* (Linnaeus)

- Colour yellow-red with black markings (Channel Islands only)
.... 1. *Hyphydrus aubei* Ganglbauer

1. *Hyphydrus aubei* Ganglbauer Plate 157

Length 4.2.9-4.6 mm. The elytral pattern is sharply defined and easily seen in the field. A Mediterranean and Central European species reaching to northern France and to the Channel Isles, where it was last recorded in 1932 in Guernsey and Jersey in quarry and heathland pools. Recorded in May, June and September.

2. *Hyphydrus ovatus* (Linnaeus) The Cherrystone Beetle Plate 158

Length 3.9-5.3 mm. Typically dark red or brown; teneral specimens do occur with a vague elytral pattern but the intensity of the pattern never approaches that of *aubei*. Females are dull with a fine elongate microreticulation barely visible at x 80. Males, though coarsely punctured, are shining. The habitat is typically deep and richly vegetated permanent lakes, ponds, ditches, canals, and occasionally river backwaters. *H. ovatus* is common across most of lowland England and Wales, including Anglesey, becoming scarce in Cumberland and north-east England. Old records for Arran and Forfar mark the extremes of this species in Scotland, which is frequent only in the south-west, and found north to the Central Belt. *H. ovatus* occurs on the Isle of Man and has a mainly central and southern distribution in Ireland. Recorded in all months, peaking in May and September.

28. *LACCORNIS* des Gozis

1. *Laccornis oblongus* (Stephens) Plate 159

Length 4.5-5.0 mm. *L. oblongus* is an elongate insect usually with a two-tone appearance, the elytra being lighter than the pronotum. In Britain this species is restricted to relic habitat in fenland, though often in very small bodies of water. It is frequent in East Anglia in "pingo fen" pools, on commons and in high quality parts of the Broads. Other English records are from Cheshire, County Durham, Herefordshire, Mid-west, North-east and South-east Yorkshire, North Somerset, Shropshire, South Lincolnshire, and Westmorland. The sole Welsh record is from Whitewall Common in Monmouthshire. In Scotland *L. oblongus* is frequent in the Borders Mosses in Roxburghshire and Selkirkshire, and at one site in Peeblesshire and two in Berwickshire with an outlier in Dumfriesshire. The northernmost record is for the Insh Marshes in East Inverness-shire. The situation in Ireland is rather different with the species frequent in many deep tussocky fens, often those that have been degraded, and it has been found in seventeen vice-counties since it was first detected in 1909. Recorded in all months, peaking in May.

Subfamily LACCOPHILINAE Gistel

29. *LACCOPHILUS* Leach

These diving beetles resemble the larger Agabini in shape. *Laccophilus* species distinguish themselves when alive by their skipping escape manoeuvres. The two common species, especially *L. hyalinus*, are greenish when alive but this fades to brown in preserved material.

Key 22. The species of *Laccophilus*

1. Hind coxae with stridulatory files (Fig. 289); green or brown with yellow markings on elytra, mainly on the edges .... 1. *Laccophilus hyalinus* (De Geer) (p. 108)

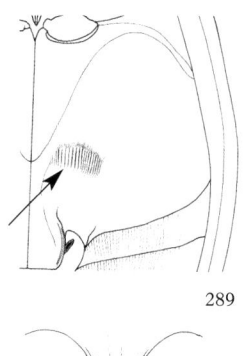

289

- Hind coxae without stridulatory files 2

   The legs should be moved aside to view in front of the hind coxal lobes. If present the stridulatory files on the hind coxae are the same size as those present on the abdominal segments of all *Laccophilus* species.

2. Elytra pale brown or greenish, with vague pale flecks; groove for the prosternal process not reaching beyond the mid coxae (Fig. 290) .... 2. *Laccophilus minutus* (Linnaeus) (p. 108)

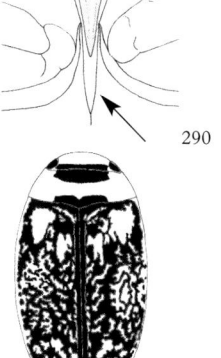

290

- Elytra generally appearing very dark, with two kinds of pattern, a series of dark, weakly defined but strongly meandering lines (resembling the feeding trails of some slugs and known as "irrorations") that coalesce in parts into a dark ground colour, superimposed upon which are yellow blotches (Fig. 291); groove for the prosternal process reaching well beyond the mid coxae, the process itself being more narrow than that of *minutus* (Fig. 292) .... 3. *Laccophilus poecilus* Klug (p. 108)

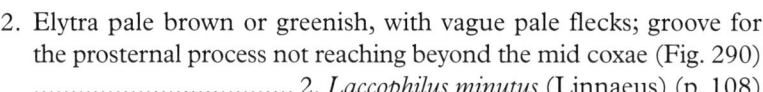

291

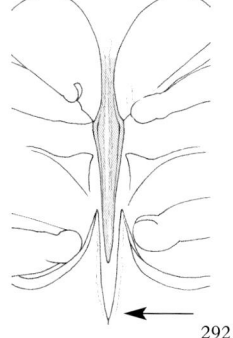

292

1. *Laccophilus hyalinus* (De Geer) Plate 160

Length 4.6-5.1 mm. The elytra, with well defined creamy yellow flecks on a greenish body colour, are usually enough to distinguish this species from the commoner *L. minutus*. This species is mainly associated with slow-moving water amongst vegetation, and it is particularly typical of canals. *L. hyalinus* occurs over most of England north to County Durham and to the south of Cumberland on the coast, being absent from Cornwall. It is scarce in Wales, mainly on the coast from Anglesey to the Gwent Levels and in the Montgomery Canal. The Scottish records, for Dumfriesshire, require confirmation. *L. hyalinus* is rare in Ireland with modern records from Kilkenny, Meath and West Meath and many earlier records. *L. hyalinus* occurs on Guernsey and Jersey. Recorded in all months, peaking in May and August.

2. *Laccophilus minutus* (**Linnaeus**) Plate 161

Length 4.3-4.8 mm. This is a common species of lowland ponds, lakes and ditches, rarely found in slow running water. It occurs across the whole of lowland Ireland, England and Wales, becoming scarcer in southern Scotland where it reaches, as an old record, to Angus. It has been reported from many islands: Arran, Bute, Cumbrae and Islay in Scotland, the Isle of Man and Anglesey, Achill, Inishbofin, Rathlin and Tory in Ireland, Lundy, Bryher, St Agnes, St Mary's, and the Isle of Wight in England, and Alderney, Guernsey and Jersey. Recorded in all months, peaking in May and August.

3. *Laccophilus poecilus* **Klug** The Puzzled Skipper Plate 162

Length 3.4-4.0 mm. Publicity associated with the conservation status of this species has generated records based on misidentification of the other *Laccophilus* species, and even *Hydroporus palustris*, which it vaguely resembles in size and colouring. The frequent changes of name to which this species has been subject have also caused confusion. As a small and mainly black *Laccophilus* this species should be easily identifiable in the field without the need to invoke the other characters in the key. *L. poecilus* occupies two quite distinct kinds of stagnant water habitat, lowland rich fen, often in grazing fen ditches near the coast but not in brackish water, and dystrophic waters. The southern distribution, from South Hampshire to East Kent, would appear to have been associated with the lowland fen habitat, with the last record in 2002 on the Lewes Levels in East Sussex. The South-west Yorkshire records, of which the last was in 1990, would appear to be associated with the peatland habitat. Recorded in February and March, and from May to November with most records in July.

Distributions

Each distribution is for Scotland (S), England (E), Wales (W), the Isle of Man (M), Northern Ireland (NI), the Republic of Ireland (RI), and the Channel Isles (CI), with records from 1980 onwards (●), before 1980 (○) and fossils (◐) since the last Ice Age.

	S	E	W	M	NI	RI	CI
GYRINIDAE							
Gyrinus aeratus Stephens	●	●	●		●	●	○
Gyrinus caspius Ménétriés	●	●	●	○	●	●	●
Gyrinus distinctus Aubé	●	●	●		●	●	
Gyrinus marinus Gyllenhal	●	●	●	●	●	●	
Gyrinus minutus Fab.	●	●	●	○	●	●	○
Gyrinus natator (L.)	○				●	●	
Gyrinus opacus Sahlberg	●						
Gyrinus paykulli Ochs	●	●	●		●	●	●
Gyrinus substriatus Stephens	●	●	●	●	●	●	●
Gyrinus suffriani Scriba	○	●	●				●
Gyrinus urinator Illiger		●	●	●	●	●	●
Orectochilus villosus (Müller)	●	●	●	●	●	●	
HALIPLIDAE							
Brychius elevatus (Panzer)	●	●	●	○	●	●	
Peltodytes caesus (Duftschmid)	●	●					
Haliplus apicalis Thomson	○	●	○		●	○	
Haliplus confinis Stephens	●	●	●	○	●	●	○
Haliplus flavicollis Sturm	●	●	●	○	●	●	●
Haliplus fluviatilis Aubé	●	●	●		●	●	
Haliplus fulvus (Fab.)	●	●	●	○	●	●	●
Haliplus furcatus Seidlitz	●						○
Haliplus heydeni Wehncke	●	●					
Haliplus immaculatus Gerhardt	●	●	●	●	●	●	○
Haliplus laminatus (Schaller)	●	●					
Haliplus lineatocollis (Marsham)	●	●	●	●	●	●	●
Haliplus lineolatus Mannerheim	●	●	●		●	●	
Haliplus mucronatus Stephens	●	●					
Haliplus obliquus (Fab.)	●	●	●	●	●		
Haliplus ruficollis (De Geer)	●	●	●	●	●	●	●
Haliplus sibiricus Motschulsky	●	●	●	●	●	●	○
Haliplus variegatus Sturm		●	●		●	●	
Haliplus varius Nicolai		●					
NOTERIDAE							
Noterus clavicornis (De Geer)	●	●	●	●	●	●	●
Noterus crassicornis (Müller)	●	●	●		●	●	
PAELOBIIDAE							
Hygrobia hermanni (Fab.)	●	●	●	●		●	●

	S	E	W	M	NI	RI	CI
DYTISCIDAE							
Agabus affinis (Paykull)	●	●	●	●	●	●	
Agabus arcticus (Paykull)	●	●	●		●	●	
Agabus biguttatus (Olivier)	●	●	●	○	●	●	
Agabus bipustulatus (L.)	●	●	●	●	●	●	●
Agabus brunneus (Fab.)		●					
Agabus congener (Thunberg)	●	●	●		●	●	
Agabus conspersus (Marsham)	●	●	●			○	●
Agabus didymus (Olivier)	○	●	●				
Agabus guttatus (Paykull)	●	●	●	●	●	●	●
Agabus labiatus (Brahm)	●	●	●	●		●	
Agabus melanarius Aubé	●	●	●		●		
Agabus nebulosus (Forster)	●	●	●	●	●	●	●
Agabus paludosus (Fab.)	●	●	●	●	●	●	●
Agabus striolatus (Gyllenhal)		●					
Agabus sturmii (Gyllenhal)	●	●	●	●	●	●	
Agabus uliginosus (L.)	●	●	●				
Agabus undulatus (Schrank)		●					
Agabus unguicularis (Thomson)	●	●	●	●	●	●	○
Ilybius aenescens Thomson	●	●	●	○	●	●	○
Ilybius ater (De Geer)	●	●	●	●	●	●	○
Ilybius chalconatus (Panzer)	●	●	●		●	●	
Ilybius fenestratus (Fab.)	●	●	●				
Ilybius fuliginosus (Fab.)	●	●	●	●	●	●	
Ilybius guttiger (Gyllenhal)	●	●	●	●	●		
Ilybius montanus (Stephens)	●	●	●	●	●	●	●
Ilybius quadriguttatus (Lacordaire)		●	●	●	●	●	●
Ilybius subaeneus Erichson	●	●			●	●	
Ilybius wasastjernae (Sahlberg)	●	◐					
Platambus maculatus (L.)	●	●	●			●	
Colymbetes fuscus (L.)	●	●	●	●	●	●	●
Rhantus bistriatus (Bergsträsser)		○					
Rhantus exsoletus (Forster)	●	●	●		●	●	●
Rhantus frontalis (Marsham)	●	●	●	●	●		
Rhantus grapii (Gyllenhal)		●	●	●	●	●	
Rhantus suturalis (Macleay)	●	●	●	●	●	●	
Rhantus suturellus (Harris)	●	●	●	●	●	●	○
Liopterus haemorrhoidalis (Fab)	●	●	●	●	●	●	
Acilius canaliculatus (Nicolai)	●	●	●	●	●	●	

	S	E	W	M	NI	RI	CI
Acilius sulcatus (L.)		●	●	●	●	●	●
Graphoderus bilineatus (De Geer)		○					
Graphoderus cinereus (L.)		●					
Graphoderus zonatus (Hoppe)		●					
Cybister lateralimarginalis (De Geer)		●					
Dytiscus circumcinctus Ahrens		●	●		●	●	
Dytiscus circumflexus Fab.	●	●	●	●		●	
Dytiscus dimidiatus Bergsträsser	●	●					
Dytiscus lapponicus Gyllenhal	●		●		●	●	
Dytiscus marginalis L.	●	●	●	●	●	●	●
Dytiscus semisulcatus Müller	●	●	●	●	●	●	○
Hydaticus seminiger (De Geer)	●	●			●	●	●
Hydaticus transversalis (Pontoppidan)				●	●		?
Bidessus minutissimus (Germar)	●	○	●	○		○	○
Bidessus unistriatus (Goeze)	●						
Hydroglyphus geminus (Fab.)			●	●			●
Deronectes latus (Stephens)	●	●	●	●			
Graptodytes bilineatus (Sturm)		●	○			●	●
Graptodytes flavipes (Olivier)		●	○				○
Graptodytes granularis (L.)	●	●	●			●	○
Graptodytes pictus (Fab.)	●	●	●	●	●	●	
Hydroporus angustatus Sturm	●	●	●	●	●	●	
Hydroporus discretus Fairmaire	●	●	●	●	●	●	
Hydroporus elongatulus Sturm	●	●					
Hydroporus erythrocephalus (L.)	●	●	●	●	●	●	
Hydroporus ferrugineus Stephens	●	●	●				
Hydroporus glabriusculus Aubé	●	●			●	●	
Hydroporus gyllenhalii Schiödte	●	●	●	●	●	●	●
Hydroporus incognitus Sharp	●	●	●	●	●	●	○
Hydroporus longicornis Sharp	●	●	●	●	○	●	
Hydroporus longulus Mulsant	●	●	●	○	●	●	
Hydroporus marginatus (Duftschmid)				●			
Hydroporus melanarius Sturm	●	●	●	●	●	●	
Hydroporus memnonius Nicolai	●	●	●	●	●	●	●
Hydroporus morio Aubé	●	●	●	●	●		
Hydroporus necopinatus Fery		●					●
Hydroporus neglectus Schaum			●	●		●	●
Hydroporus nigrita (Fab.)	●	●	●	●	●	●	
Hydroporus obscurus Sturm	●	●	●	○	●	●	
Hydroporus obsoletus Aubé	●	●	●	●	●	●	
Hydroporus palustris (L.)	●	●	●	●	●	●	
Hydroporus planus (Fab.)	●	●	●	●	●	●	
Hydroporus pubescens (Gyllenhal)	●	●	●	●	●	●	

	S	E	W	M	NI	RI	CI
Hydroporus rufifrons (Müller)	●	●	●				
Hydroporus scalesianus Stephens	●	●	●		●	●	
Hydroporus striola (Gyllenhal)	●	●	●	●	●	●	○
Hydroporus tessellatus Drapiez	●	●	●	●	●	●	●
Hydroporus tristis (Paykull)	●	●	●	●	●	●	
Hydroporus umbrosus (Gyllenhal)	●	●	●	●	●	●	
Nebrioporus assimilis (Paykull)	●	●	●	●	●	●	
Nebrioporus canaliculatus (Lacordaire)				●			
Nebrioporus depressus (Fab.)	●	●		?		●	●
Nebrioporus elegans (Panzer)	●	●	●	●		●	●
Oreodytes alpinus (Paykull)	●						
Oreodytes davisii (Curtis)	●	●	●				
Oreodytes sanmarkii (Sahlberg)	●	●	●			●	
Oreodytes septentrionalis (Gyllenhal)	●	●	●	●	●	●	
Porhydrus lineatus (Fab.)	●	●	●	●	●		○
Scarodytes halensis (Fab.)	●						
Suphrodytes dorsalis (Fab.)	●				●	●	
Suphrodytes figuratus (Gyllenhal)	●				●	●	
Stictonectes lepidus (Olivier)	●	●	●	●	●	●	○
Stictotarsus duodecimpustulatus (Fab.)	●	●	●	●	●	●	●
Boreonectes multilineatus (Falkenström)	●	●	●	●	●	●	
Hydrovatus clypealis Sharp			●	●		●	●
Hydrovatus cuspidatus (Kunze)			●				
Hygrotus confluens (Fab.)	●	●	●				
Hygrotus decoratus (Gyllenhal)	●	●				●	
Hygrotus impressopunctatus (Schaller)	●	●	●	●	●	●	●
Hygrotus inaequalis (Fab.)	●	●	●	●	●	●	●
Hygrotus nigrolineatus (von Steven)	●	●	●				
Hygrotus novemlineatus (Stephens)	●	●			●	●	
Hygrotus parallellogrammus (Ahrens)			●	●		?	?
Hygrotus quinquelineatus (Zetterstedt)	●	●			●	●	
Hygrotus versicolor (Schaller)	●	●	●			●	
Hyphydrus aubei Ganglbauer							○
Hyphydrus ovatus (L.)		●	●	●	●	●	●
Laccornis oblongus (Stephens)	●	●	●		●	●	
Laccophilus hyalinus (De Geer)	?	●	●		○	●	●
Laccophilus minutus (L.)	●	●	●	●	●	●	●
Laccophilus poecilus Klug		●					

References

Angus, R.B. (2010). *Boreonectes* gen. n., a new genus for the *Stictotarsus griseostriatus* (De Geer) group of sibling species (Coleoptera: Dytiscidae), with additional karyosystematic data on the group. *Comparative Cytogenetics* **4** (2): 123-131.

Balfour-Browne, W.A.F. (1940). *British Water Beetles*. **1**, Ray Society, London.

Balfour-Browne, W.A.F. (1950). *British Water Beetles*. **2**, Ray Society, London.

Balfour-Browne, W.A.F. (1953). Coleoptera: Hydradephaga. *Handbooks for the identification of British insects* **4** (3): 1-34. Royal Entomological Society, London.

Balfour-Browne, W.A.F. (1958). *British Water Beetles*. **3**, Ray Society, London.

Balke, M., Ribera, I., Hendrich, L. Miller, M.A., Sagata, K., Posman, A., Vogler, A.P. & Meier, R. (2009). New Guinea highland origin of a widespread arthropod supertramp. *Proceedings of the Royal Society Series* B **276**: 2359-2367 doi:10.1098/rspb.2009.0015.

Balke, M., Ribera, I. & Vogler, A.P. (2004). MtDNA phylogeny and biogeography of Copelatinae, a highly diverse group of tropical diving beetles (Dytiscidae). *Molecular Phylogenetics and Evolution* **32**: 866-880.

Bilton, D.T., Thompson, A. & Foster, G.N. (2008). Inter- and intrasexual dimorphism in the diving beetle *Hydroporus memnonius* Nicolai (Coleoptera: Dytiscidae). *Journal of the Linnean Society* **94**: 685-697.

Fery, H. (1999). Revision of a part of the *memnonius*-group of *Hydroporus* Clairville, 1806 (Insecta: Coleoptera: Dytiscidae). *Annalen der Naturhistorischen Museums in Wien* **101 B**: 217-269.

Foster, G.N. (2004). An annotated checklist of British and Irish water beetles, and associated taxa: Myxophaga and Adephaga – Hydradephaga. *The Coleopterist* **13**: 14-160.

Foster, G. [N.] (2008). Whirligigs in Britain and Ireland. *British Wildlife*, October 2008 28-35.

Foster, G.N. (2010). *A review of the scarce and threatened Coleoptera of Great Britain. Part 3: water beetles.* Species Status No. 1. Joint Nature Conservation Committee, Peterborough.

Foster, G.N., Bilton, D.T., Routledge, S. & Eyre, M.D. (2008). The past and present statuses of *Hydroporus rufifrons* (Müller) (Coleoptera, Dytiscidae) in Great Britain. *The Coleopterist* **17**: 51-63.

Foster, G.N. & Carr, R. (2008). The status of *Bidessus unistriatus* (Goeze) in England, with records of *B. grossepunctatus* Vorbringer, 1907, a species present in England in the Bronze Age. *The Coleopterist* **17**: 191-203.

Foster, G.N., Nelson, B.H. & O Connor, Á. (2009). *Ireland Red List No. 1. Water Beetles*. National Parks & Wildlife, Department of Environment, Heritage & Local Government, Dublin.

Friday, L.E. (1988). A key to the adults of British water beetles. *Field Studies* **7**: 1-151. Published separately as *AIDGAP Book* **189**, Field Studies Council, Taunton.

Kelly, K.B. & Nilsson, A.N. (2003). Homology and terminology: communicating information about rotated structures in water beetles. *Latissimus* **17**: 1-3.

Löbl, I. & Šmetana, A. (Eds) (2003). *Catalogue of Palaearctic Coleoptera, Volume 1 Archostemata – Myxophaga – Adephaga*. Apollo Books, Stenstrup.

Luff, M.L. (2007). The Carabidae (ground beetles) of Britain and Ireland. *Handbooks for the Identification of British Insects* **4** (2) 2nd edition. Royal Entomological Society, St. Albans.

Mann, D.J. (2006). *Ptilodactyla exotica* Chapin 1927 (Coleoptera: Ptilodactylidae: Ptilodactylinae) established breeding under glass in Britain, with a brief discussion on the family Ptilodactylidae. *Entomologist's Monthly Magazine* **142**: 67-79.

Morris, M.G. (2002). True weevils (Part I). Coleoptera: Curculionidae (Subfamilies Raymondionyminae to Smicronychinae). *Handbooks for the Identification of British Insects* **5** (17b). Royal Entomological Society, London.

Morris, M.G, (2008). True weevils (Part II) (Coleoptera: Curculionidae, Ceutorhynchinae). *Handbooks for the identification of British insects* Volume 5, Part 17c. Royal Entomological Society, St. Albans.

Nilsson, A. N. (2000). A new view of the generic classification of the *Agabus*-group of genera of the Agabini, aimed at solving the problem with a paraphyletic *Agabus* (Coleoptera: Dytiscidae). *Koleopterologische Rundschau* **70**:17-36.

Nilsson, A.N. & van Vondel, B.J. (2005). *World Catalogue of Insects.* Volume 7, *Amphizoidae, Aspidytidae, Haliplidae, Noteridae and Paelobiidae (Coleoptera, Adephaga).* Apollo Books Aps., Stenstrup.

Scheffer, M. & Nes, E.H. van (2006). Self-organized similarity, the evolutionary emergence of groups of similar species. *Proceedings of the National Academy of Sciences of the United States of America* **103**: 6230-6235.

Sharp, D. & Muir, F. (1912). The comparative anatomy of the male genital tube in Coleoptera. *Transactions of the Entomological Society of London* **60**: 477-641 + plates xlii-lxxviii.

Shirt, D.B. (1983). Studies on the *Potamonectes depressus* (Fabricius) complex of aquatic Coleoptera (Dytiscidae). Unpublished Ph.D. thesis, Royal Holloway College, University of London.

Toledo, M. (2009). Revision in part of the genus *Nebrioporus* Régimbart, 1906, with emphasis on the *N. laeviventris*-group (Coleoptera: Dytiscidae). *Zootaxa* **2040:** 1-111.

Appendix
Guidance on water beetles for non-specialists

Before going further make sure that the specimen to be identified really is a beetle. Aquatic bugs such as water boatmen can be mistaken for water beetles as they too have the front pair of wings modified into wing-cases. The most obvious difference between them and beetles is that these wing-cases overlap whereas they meet in the midline (or suture) in beetles. The overlap produces a diagonal cross (Fig. 293). Bugs have "incomplete metamorphosis", i.e. the immature stages resemble small versions of the adult, with the wings absent in the first instars and then appearing as wing buds after a few moults, so a beetle-shaped animal without elytra is likely to be an immature bug.

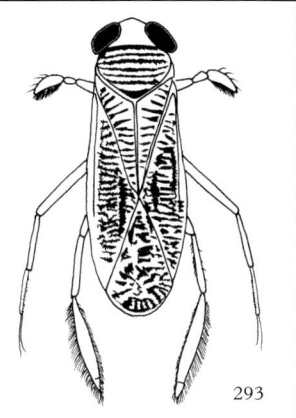

293

Key 23. The families of aquatic beetles

There is neither a single character nor a set of characters for recognising an aquatic beetle. Most beetles likely to occur in net samples are keyed below, which is intended mainly to help limnologists get to the right part of species identification keys. If your specimen will not key here, it has probably fallen in!

Several terrestrial families of beetle have a few wetland representatives not included in this key. For example, several ground beetles are confined to shorelines and to wet shingle, with some even found in the tidal zone in rock crevices. These with the truly aquatic *Oodes helopioides* (Fab.) are keyed by Luff (2007). A few ladybird beetles (Coccinellidae) are confined to wetland, in particular the *Coccidula* species and the nineteen-spot ladybird *Anisosticta novemdecimpunctata* (L.). The rove beetles (Staphylinidae) have many species associated with wet litter, and are beyond the scope of the current work. By contrast a largely aquatic family, the Ptilodactylidae, is represented by an introduced species confined to moist soil in greenhouses in England (Mann, 2006).

1. A rare and exceptionally small beetle, 1.2 mm at most, black and globular (Fig. 294), living on moist soil beside water ....
.... **SPHAERIUSIDAE** (suborder MYXOPHAGA) (Part II)

294

- All other beetles; if small, black and globular, then not less than 1.5 mm long .... Suborder ADEPHAGA and POLYPHAGA

2

There are other minute black beetles, some aquatic and some terrestrial, none of them being entirely globular with the elytra completely covering the abdomen.

2. Middle and hind legs shorter than the front legs and very broad (Fig. 295); eyes divided horizontally to produce two pairs; can swim on the surface of the water; size range 3.5-7.8 mm .... **GYRINIDAE** whirligig beetles (p. 12)

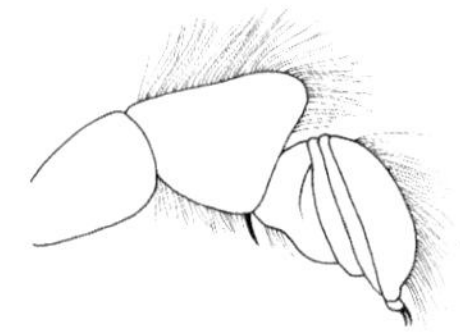

295

- Middle and hind legs as long as the front legs; eyes undivided; swim below the surface or float on it; size range 1-48 mm 3

3. Antennae long and thread-like throughout most of their length, with 7-11 segments visible (Fig. 296), sometimes with the middle segments broader than the rest (Fig. 297) but never with a club at the end 4

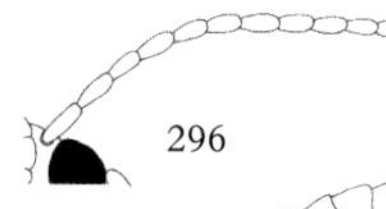

296

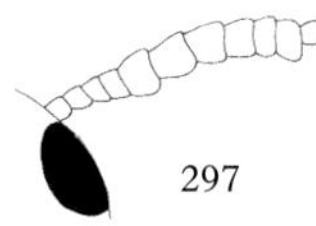

297

- Antennae short with a terminal club (Fig. 298), or short with one or two basal segments greatly much larger than the rest (Fig. 299). If the antennae cannot be seen apply a fine brush to grooves on the front of the head in which they may be tucked 11

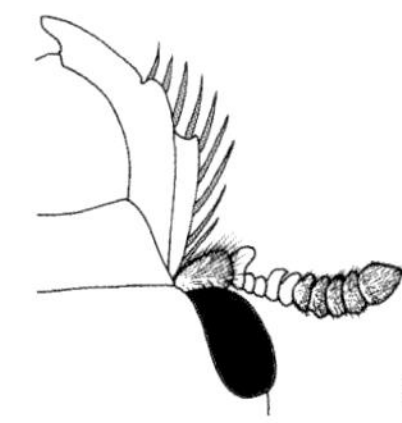

298

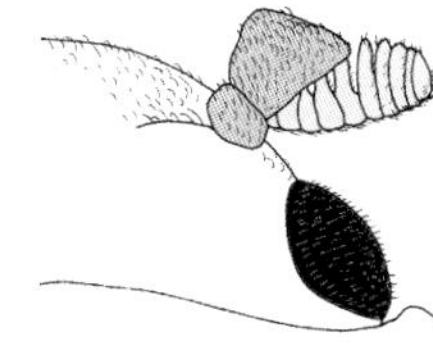

299

4. Antennae inserted on the front of the head, the distance between their bases the same as or less than the length of their first segment (Fig. 300); most species found above the water, either more than 4 mm long with a metallic finish to the head, thorax and elytra or about 2 mm long and brown 5

- Antennae inserted further apart than the length of the first segment; metallic colouring if present limited to parts of the head and pronotum, less often the elytra, or with a brassy finish to an otherwise black beetle; 1-38 mm long; species found above or below the water 6

300

5. Large, elongated beetles with the head prominent; elytra with at least five striae made up of rows of large punctures running between the centre and the shoulder, the additional stria next to the suture often being a short series of punctures on the front third of the elytron; species found above the water with metallic reflections; much rarer fully aquatic species with yellow and black or brown stripes; 5-11 mm .... **CHRYSOMELIDAE** Donaciinae or reed beetles (Part II)

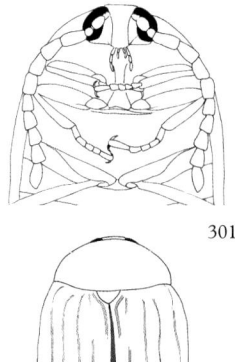

- Small, rounded beetle with head tucked in so far that it may be invisible from above (Fig. 301); elytra with four striae as deep grooves between the centre and the shoulder on each elytron in addition to the sutural stria, which is usually confined to the front half of the elytron (Fig. 302); 1.8-2.1 mm .... **PSEPHENIDAE** water penny beetle (Part II)

301

302

6. Fourth segment of hind tarsus bilobed with the claw-bearing fifth segment rising from it (Fig. 303); elytra soft and well covered with hair; 2.4 – 5.5 mm .... **SCIRTIDAE** marsh beetles (Part II)

- Fourth segment of hind tarsus like the other non-claw bearing segments, not bilobed; elytra hard in mature beetles and with hair limited, usually on punctures in rows, or almost completely absent 7

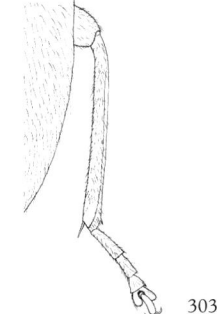

303

7. Pronotum as long as wide; head and eyes almost hidden below the convex front margin of the pronotum (Fig. 304) (beware specimens preserved in alcohol where the head may protrude artificially); hind tarsi with the last segment very long and bulbous at tip (Fig. 305), not hairy; size range 1.3-4.8 mm .... **ELMIDAE** riffle beetles (Part II)

304

- Pronotum wider than long; head and eyes visible from above; last segment of hind tarsi not very long or as bulbous, usually with a fringe of long swimming hairs; size range 1.7-38.0 mm 8

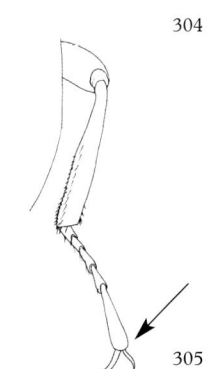

305

8. Hind coxae with large, rounded plates covering half of the abdomen and the basal half of the hind femora (Fig. 306); elytra each with about 10 longitudinal rows of large punctures, at least five of them on the dorsal side inside the shoulder; size range 2.5-5.0 mm .... **HALIPLIDAE** crawling water beetles (p. 20)

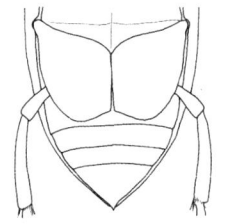

306

- Hind coxae with lobed or pointed projections, most of the hind femora visible (Fig. 307); elytra each with not more than 5 longitudinal lines in total, usually made up of small punctures and/or grooves; size range 1.7-38.0 mm 9

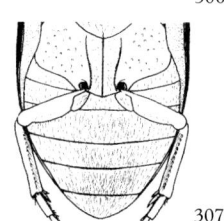

307

9. Head narrower than the front of the pronotum; beetle 8.5-10 mm long, strongly convex below; elytra black with yellow or red front and side margins (Fig. 308); loudly stridulates when alarmed .... **PAELOBIIDAE** squeak beetle (p. 31)

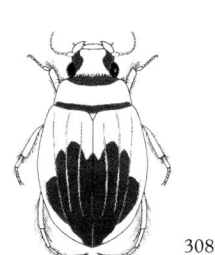

- Head about the same width as the front of the pronotum and its outline usually forming a smooth curve with that of the pronotum (beware specimens preserved in alcohol where the head may be extended beyond its normal position); differing from the squeak beetle in the combination of colour and convexity 10

308

10. Hind coxal processes very broad (together at least as wide as long) forming a distinctive plate, the "noterid platform" (shaded in Fig. 309); middle antennal segments expanded (Fig. 310); size 3.5-5.0 mm ...... **NOTERIDAE** burrowing diving beetles (p. 30)

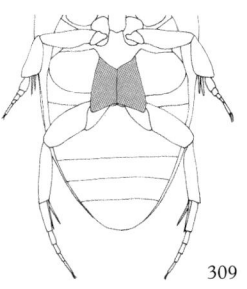

309

310

- Hind coxal processes longer than broad, hind margin variously shaped (Fig. 311) but not as in the noterid platform; middle antennal segments usually longer than wide and thin; size 1.7-38.0 mm .... **DYTISCIDAE** diving beetles (p. 31)

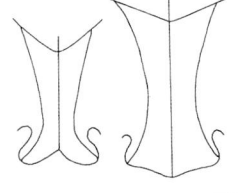

311

11. Head with a snout or proboscis, at least as long as broad (examine from the side as well as from above); antennae long, with a definite "elbow" between the first and second segment in most species (Fig. 312) .... **CURCULIONIDAE** weevils
– consult Morris (2002, 2008)

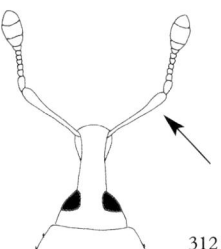

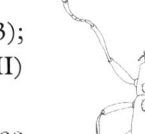

- No such snout, though the front of the head may bulge out in the centre; antennae without an "elbow" 12

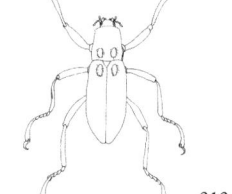
312

12. Front and middle legs longer than the entire beetle (Fig. 313); antennae short .... **ELMIDAE** riffle beetles (Part II)

- Front and middle legs shorter than the entire beetle; antennae long or short 13

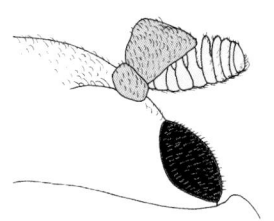
313

13. Elytra and pronotum covered with dense hair; sizes 2.5-6.2 mm 14

- Elytra and pronotum without hairs, or with sparse hairs; sizes 1-48 mm 15

14. Antennae short, with an expanded second segment, often tucked into the front of the head (Fig. 314); last segment of the tarsi long; all tibiae slender; elytra uniformly greyish brown; 3.8-5.4 mm .... **DRYOPIDAE** (Part II)

314

- Antennae with first segment large, most of the rest forming a narrow club (Fig. 315); all tarsal segments short; front tibiae very broad; elytra patterned black or brown with yellow, coalescing flecks; 2.5-6.2 mm .... **HETEROCERIDAE** (Part II)

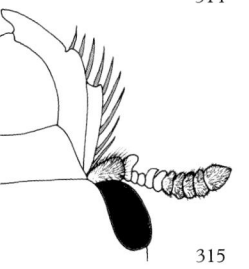
315

15. Palps from at least two-thirds to 4 times as long as the antennae and easily seen (Figs 316 and 317); antennae often hidden and each with a club with 3 or 5 enlarged hairy segments; 1-48 mm .... **HYDROPHILOIDEA** and **HYDRAENIDAE** (Part II)
"palpicorn" beetles

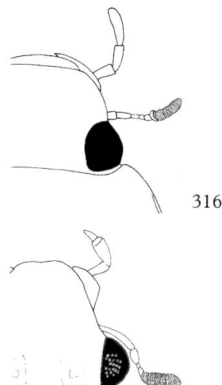

316

317

- Palps short, much less than two-thirds as long as the antennae, and not usually visible; antennae longer than the head and pronotum, with one segment in the club (Fig. 318); beetle convex below, black with yellow bases of the antennae and legs; size 1.5-1.8 mm .... **LIMNICHIDAE** (Part II)

View from in front as well as from above.

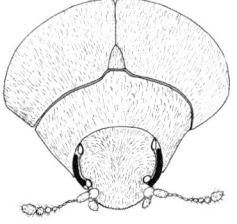

318

Index

Main entries and start of main sections are shown in **bold**. Synonyms are given in *italics*.

Colour plates

Plate 1. Family Gyrinidae
Gyrinus minutus Fabricius
3.5-4.5 mm (page 17)

Plate 2. Family Gyrinidae
Gyrinus aeratus Stephens
4.5-6.3 mm (page 17)

Plate 3. Family Gyrinidae
Gyrinus caspius Ménétriés
5.0-7.5 mm (page 17)

Plate 4. Family Gyrinidae
Gyrinus distinctus Aubé
5.0-7.0 mm (page 18)

Plate 5. Family Gyrinidae
Gyrinus marinus Gyllenhal
4.5-7.5 mm (page 18)

Plate 6. Family Gyrinidae
Gyrinus natator (Linnaeus)
4.5-6.1 mm (page 18)

Plate 7. Family Gyrinidae
Gyrinus opacus Sahlberg
5.0-6.5 mm (page 19)

Plate 8. Family Gyrinidae
Gyrinus paykulli Ochs
5.5-7.8 mm (page 19)

Plate 9. Family Gyrinidae
Gyrinus substriatus Stephens
5.0-7.0 mm (page 19)

Plate 10. Family Gyrinidae
Gyrinus suffriani Scriba
4.0-6.2 mm (page 19)

Plate 11. Family Gyrinidae
Gyrinus urinator Illiger
5.0-7.8 mm (page 20)

Plate 12. Family Gyrinidae
Orectochilus villosus (O.F. Müller)
5.5-6.5 mm (page 20)

Plate 13. Family Haliplidae
Brychius elevatus (Panzer)
3.5-4.4 mm (page 22)

Plate 14. Family Haliplidae
Haliplus confinis Stephens
3.0-3.5 mm (page 26)

Plate 15. Family Haliplidae
Haliplus obliquus (Fabricius)
3.0-3.5 mm (page 26)

Plate 16. Family Haliplidae
Haliplus varius Nicolai
2.6-2.9 mm (page 26)

Plate 17. Family Haliplidae
Haliplus apicalis Thomson
2.5-3.0 mm (page 27)

Plate 18. Family Haliplidae
Haliplus fluviatilis Aubé
2.5-2.8 mm (page 27)

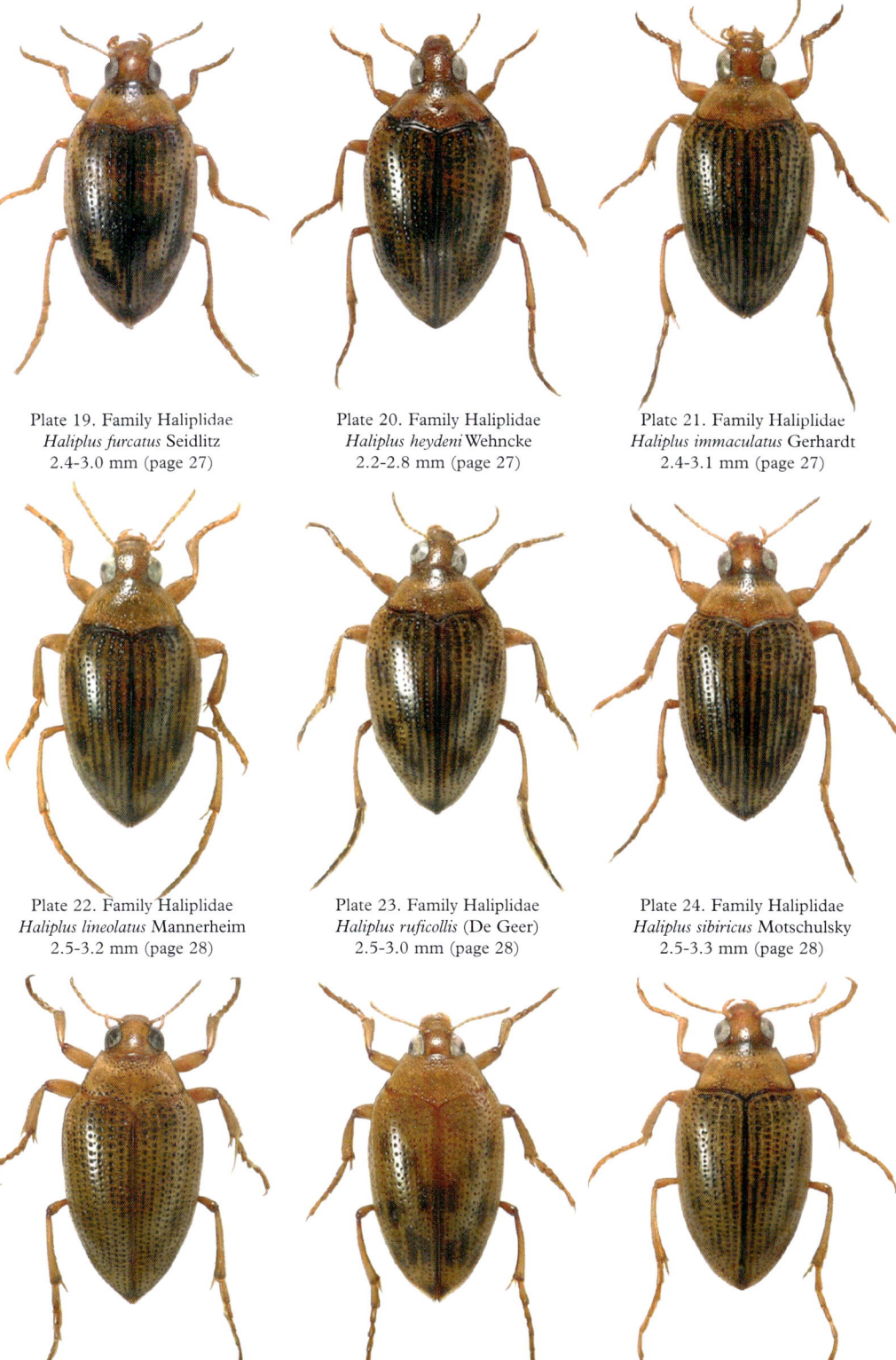

Plate 19. Family Haliplidae
Haliplus furcatus Seidlitz
2.4-3.0 mm (page 27)

Plate 20. Family Haliplidae
Haliplus heydeni Wehncke
2.2-2.8 mm (page 27)

Platc 21. Family Haliplidae
Haliplus immaculatus Gerhardt
2.4-3.1 mm (page 27)

Plate 22. Family Haliplidae
Haliplus lineolatus Mannerheim
2.5-3.2 mm (page 28)

Plate 23. Family Haliplidae
Haliplus ruficollis (De Geer)
2.5-3.0 mm (page 28)

Plate 24. Family Haliplidae
Haliplus sibiricus Motschulsky
2.5-3.3 mm (page 28)

Plate 25. Family Haliplidae
Haliplus flavicollis Sturm
3.5-4.0 mm (page 28)

Plate 26. Family Haliplidae
Haliplus fulvus (Fabricius)
3.6-4.2 mm (page 28)

Plate 27. Family Haliplidae
Haliplus laminatus (Schaller)
2.5-3.0 mm (page 29)

Plate 28. Family Haliplidae
Haliplus mucronatus Stephens
4.0-5.0 mm (page 29)

Plate 29. Family Haliplidae
Haliplus variegatus Sturm
2.5-3.5 mm (page 29)

Plate 30. Family Haliplidae
Haliplus lineatocollis (Marsham)
2.6-3.5 mm (page 29)

Plate 31. Family Haliplidae
Peltodytes caesus (Duftschmid)
3.5-4.0 mm (page 30)

Plate 32. Family Noteridae
Noterus clavicornis (De Geer)
4.0-5.0 mm (page 30)

Plate 33. Family Noteridae
Noterus crassicornis (Müller)
3.5-4.0 mm (page 31)

Plate 34. Family Paelobiidae
Hygrobia hermanni (Fabricius)
8.5-10 mm (page 31)

Plate 35. Family Dytiscidae
Agabus arcticus (Paykull)
6.7-8.3 mm (page 50)

Plate 36. Family Dytiscidae
Agabus congener (Thunberg)
6.6-8.0 mm (page 50)

Plate 37. Family Dytiscidae
Agabus sturmii (Gyllenhal)
7.7-8.9 mm (page 50)

Plate 38. Family Dytiscidae
Agabus labiatus (Brahm)
5.8-6.8 mm (page 51)

Plate 39. Family Dytiscidae
Agabus uliginosus (Linnaeus)
6.4-7.6 mm (page 51)

Plate 40. Family Dytiscidae
Agabus uliginosus dispar Bold
6.4-7.6 mm (page 51)

Plate 41. Family Dytiscidae
Agabus undulatus (Schrank)
7.0-7.8 mm (page 51)

Plate 42. Family Dytiscidae
Agabus affinis (Paykull)
6.2-7.1 mm (page 52)

Plate 43. Family Dytiscidae
Agabus biguttatus (Olivier)
8.4-9.0 mm (page 52)

Plate 44. Family Dytiscidae
Agabus bipustulatus (Linnaeus)
9.5-11.6 mm (page 52)

Plate 45. Family Dytiscidae
Agabus brunneus (Fabricius)
7.6-9.1 mm (page 53)

Plate 46. Family Dytiscidae
Agabus conspersus (Marsham)
7.0-8.3 mm (page 53)

Plate 47. Family Dytiscidae
Agabus didymus (Olivier)
7.5-8.0 mm (page 53)

Plate 48. Family Dytiscidae
Agabus guttatus (Paykull)
7.8-9.2 mm (page 53)

Plate 49. Family Dytiscidae
Agabus melanarius Aubé
8.5-9.8 mm (page 53)

Plate 50. Family Dytiscidae
Agabus nebulosus (Forster)
8.2-8.6 mm (page 54)

Plate 51. Family Dytiscidae
Agabus paludosus (Fabricius)
6.5-8.0 mm (page 54)

Plate 52. Family Dytiscidae
Agabus striolatus (Gyllenhal)
7.2-7.9 mm (page 54)

Plate 53. Family Dytiscidae
Agabus unguicularis (Thomson)
6.0-6.7 mm (page 54)

Plate 54. Family Dytiscidae
Ilybius aenescens Thomson
8.5-9.8 mm (page 55)

Plate 55. Family Dytiscidae
Ilybius ater (De Geer)
12.5-14.5 mm (page 55)

Plate 56. Family Dytiscidae
Ilybius chalconatus (Panzer)
7.5-8.7 mm (page 55)

Plate 57. Family Dytiscidae
Ilybius fenestratus (Fabricius)
10.0-12.0 mm (page 56)

Plate 58. Family Dytiscidae
Ilybius fuliginosus (Fabricius)
10.0-11.5 mm (page 56)

Plate 59. Family Dytiscidae
Ilybius guttiger (Gyllenhal)
8.7-10.0 mm (page 56)

Plate 60. Family Dytiscidae
Ilybius montanus (Stephens)
6.9-8.5 mm (page 56)

Plate 61. Family Dytiscidae
Ilybius quadriguttatus (Lacordaire)
10.5-12.2 mm (page 57)

Plate 62. Family Dytiscidae
Ilybius subaeneus Erichson
10.0-12.5 mm (page 57)

Plate 63. Family Dytiscidae
Ilybius wasastjernae (Sahlberg)
6.4-7.3 mm (page 57)

Plate 64. Family Dytiscidae
Platambus maculatus (Linnaeus)
7.5-8.5 mm (page 57)

Plate 65. Family Dytiscidae
Colymbetes fuscus (Linnaeus)
15.0-17.0 mm (page 58)

Plate 66. Family Dytiscidae
Rhantus grapii (Gyllenhal)
10.0-11.0 mm (page 60)

Plate 67. Family Dytiscidae
Rhantus bistriatus (Bergsträsser)
9.0-9.5 mm (page 60)

Plate 68. Family Dytiscidae
Rhantus exsoletus (Forster)
9.0-10.0 mm (page 60)

Plate 69. Family Dytiscidae
Rhantus frontalis (Marsham)
9.4-11.4 mm (page 60)

Plate 70. Family Dytiscidae
Rhantus suturalis (Macleay)
10.0-13.0 mm (page 61)

Plate 71. Family Dytiscidae
Rhantus suturellus (Harris)
9.4-11.0 mm (page 61)

Plate 72. Family Dytiscidae
Liopterus haemorrhoidalis (Fabricius)
6.3-7.9 mm (page 61)

Plate 73. Family Dytiscidae
Acilius canaliculatus (Nicolai) ♂
14.0-15.5 mm (page 62)

Plate 74. Family Dytiscidae
Acilius canaliculatus (Nicolai) ♀
14.0-15.5 mm (page 62)

Plate 75. Family Dytiscidae
Acilius sulcatus (Linnaeus) ♂
15.7-18.0 mm (page 63)

Plate 76. Family Dytiscidae
Graphoderus bilineatus (De Geer) ♂
14.0-15.7 mm (page 64)

Plate 77. Family Dytiscidae
Graphoderus cinereus (Linnaeus) ♂
13.8-15.3 mm (page 64)

Plate 78. Family Dytiscidae
Graphoderus zonatus (Hoppe) ♂
12.0-15.6 mm (page 64)

Plate 79. Family Dytiscidae
Cybister lateralimarginalis (De Geer) ♂
29.0-37.0 mm (page 65)

Plate 80. Family Dytiscidae
Dytiscus circumcinctus Ahrens ♂
27.0-32.0 mm (page 66)

Plate 81. Family Dytiscidae
Dytiscus circumcinctus Ahrens male-like ♀
27.0-32.0 mm (page 66)

Plate 82. Family Dytiscidae
Dytiscus circumcinctus Ahrens sulcate ♀
27.0-32.0 mm (page 66)

Plate 83. Family Dytiscidae
Dytiscus circumflexus Fabricius ♂
26.0-32.0 mm (page 66)

Plate 84. Family Dytiscidae
Dytiscus dimidiatus Bergsträsser ♂
32.0-38.0 mm (page 67)

Plate 85. Family Dytiscidae
Dytiscus dimidiatus Bergsträsser ♀
32.0-38.0 mm (page 67)

Plate 86. Family Dytiscidae
Dytiscus lapponicus Gyllenhal ♂
22.0-28.0 mm (page 67)

Plate 87. Family Dytiscidae
Dytiscus marginalis Linnaeus ♂
26.0-32.0 mm (page 67)

Plate 88. Family Dytiscidae
Dytiscus marginalis Linnaeus ♀
26.0-32.0 mm (page 67)

Plate 89. Family Dytiscidae
Dytiscus semisulcatus Müller ♂
22.0-30.0 mm (page 67)

Plate 90. Family Dytiscidae
Dytiscus semisulcatus Müller ♀
22.0-30.0 mm (page 67)

Plate 91. Family Dytiscidae
Hydaticus seminiger (De Geer)
13.0-14.5 mm (page 68)

Plate 92. Family Dytiscidae
Hydaticus transversalis (Pontoppidan)
12.0-13.0 mm (page 68)

Plate 93. Family Dytiscidae
Bidessus minutissimus (Germar)
1.5-1.9 mm (page 69)

Plate 94. Family Dytiscidae
Bidessus unistriatus (Goeze)
1.7-2.0 mm (page 69)

Plate 95. Family Dytiscidae
Hydroglyphus geminus (Fabricius)
1.9-2.2 mm (page 69)

Plate 96. Family Dytiscidae
Deronectes latus (Stephens)
4.2-4.8mm (page 70)

Plate 97. Family Dytiscidae
Graptodytes bilineatus (Sturm)
2.3-2.7 mm (page 71)

Plate 98. Family Dytiscidae
Graptodytes flavipes (Olivier)
2.4-2.7 mm (page 71)

Plate 99. Family Dytiscidae
Graptodytes granularis (Linnaeus)
2.0-2.3 mm (page 72)

Plate 100. Family Dytiscidae
Graptodytes pictus (Fabricius)
2.0-2.4 mm (page 72)

Plate 101. Family Dytiscidac
Hydroporus angustatus Sturm
2.8-3.2 mm (page 82)

Plate 102. Family Dytiscidae
Hydroporus discretus Fairmaire &
Brisout de Barneville
3.0-3.4 mm (page 82)

Plate 103. Family Dytiscidae
Hydroporus elongatulus Sturm
3.3-3.6 mm (page 83)

Plate 104. Family Dytiscidae
Hydroporus erythrocephalus (Linnaeus)
3.4-4.3 mm (page 83)

Plate 105. Family Dytiscidae
Hydroporus ferrugineus Stephens
3.5-4.2 mm (page 83)

Plate 106. Family Dytiscidae
Hydroporus glabriusculus Aubé
2.8-3.2 mm (page 83)

Plate 107. Family Dytiscidae
Hydroporus gyllenhalii Schiødte
3.4-4.1 mm (page 84)

Plate 108. Family Dytiscidae
Hydroporus incognitus Sharp
2.9-3.9 mm (page 84)

Plate 109. Family Dytiscidae
Hydroporus longicornis Sharp
3.5-3.8 mm (page 84)

Plate 110. Family Dytiscidae
Hydroporus longulus Mulsant & Rey
3.4-3.8 mm (page 84)

Plate 111. Family Dytiscidae
Hydroporus marginatus (Duftschmid)
3.5-4.5 mm (page 85)

Plate 112. Family Dytiscidae
Hydroporus melanarius Sturm
3.0-3.6 mm (page 85)

Plate 113. Family Dytiscidae
Hydroporus memnonius Nicolai
3.8-4.3 mm (page 85)

Plate 114. Family Dytiscidae
Hydroporus memnonius castaneus Aubé
3.8-4.3 mm (page 85)

Plate 115. Family Dytiscidae
Hydroporus morio Aubé
3.0-3.7 mm (page 86)

Plate 116. Family Dytiscidae
Hydroporus necopinatus roni Fery
3.2-3.5 mm (page 86)

Plate 117. Family Dytiscidae
Hydroporus neglectus Schaum
2.3-2.7 mm (page 86)

Plate 118. Family Dytiscidae
Hydroporus nigrita (Fabricius)
2.8-3.3 mm (page 87)

Plate 119. Family Dytiscidae
Hydroporus obscurus Sturm
2.5-2.9 mm (page 87)

Plate 120. Family Dytiscidae
Hydroporus obsoletus Aubé
3.3-4.2 mm (page 87)

Plate 121. Family Dytiscidae
Hydroporus palustris (Linnaeus)
3.3-4.0 mm (page 88)

Plate 122. Family Dytiscidae
Hydroporus planus (Fabricius)
3.8-4.8 mm (page 88)

Plate 123. Family Dytiscidae
Hydroporus pubescens (Gyllenhal)
3.2-3.8 mm (page 88)

Plate 124. Family Dytiscidae
Hydroporus rufifrons (Müller)
4.2-5.3 mm (page 89)

Plate 125. Family Dytiscidae
Hydroporus scalesianus Stephens
1.9-2.2 mm (page 89)

Plate 126. Family Dytiscidae
Hydroporus striola (Gyllenhal
3.0-3.4 mm (page 89)

Plate 127. Family Dytiscidae
Hydroporus tessellatus (Drapiez)
3.4-3.8 mm (page 89)

Plate 128. Family Dytiscidae
Hydroporus tristis (Paykull)
2.8-3.3 mm (page 90)

Plate 129. Family Dytiscidae
Hydroporus umbrosus (Gyllenhal)
2.5-2.8 mm (page 90)

Plate 130. Family Dytiscidae
Nebrioporus assimilis (Paykull)
4.0-4.3 mm (page 92)

Plate 131. Family Dytiscidae
Nebrioporus canaliculatus (Lacordaire)
4.8-5.8 mm (page 92)

Plate 132. Family Dytiscidae
Nebrioporus depressus (Fabricius)
4.5-5.2 mm (page 92)

Plate 133. Family Dytiscidae
Nebrioporus elegans (Panzer)
4.4-5.0 mm (page 93)

Plate 134. Family Dytiscidae
Oreodytes alpinus (Paykull)
4.2-5.0 mm (page 94)

Plate 135. Family Dytiscidae
Oreodytes davisii (Curtis)
3.8-4.5 mm (page 94)

Plate 136. Family Dytiscidae
Oreodytes sanmarkii (Sahlberg)
2.9-3.3 mm (page 95)

Plate 137. Family Dytiscidae
Oreodytes septentrionalis (Gyllenhal)
3.2-3.6 mm (page 95)

Plate 138. Family Dytiscidae
Porhydrus lineatus (Fabricius)
3.0-3.5 mm (page 95)

Plate 139. Family Dytiscidae
Scarodytes halensis (Fabricius)
3.8-4.3 mm (page 96)

Plate 140. Family Dytiscidae
Stictonectes lepidus (Olivier)
3.1-3.4 mm (page 96)

Plate 141. Family Dytiscidae
Stictotarsus duodecimpustulatus (Fabricius)
5.2-5.7 mm (page 96)

Plate 142. Family Dytiscidae
Boreonectes multilineatus (Falkenström)
4.0-4.8 mm (page 97)

Plate 143. Family Dytiscidae
Suphrodytes dorsalis (Fabricius)
4.6-5.3 mm (page 98)

Plate 144. Family Dytiscidae
Suphrodytes figuratus (Gyllenhal)
4.3-4.6 mm (page 99)

Plate 145. Family Dytiscidae
Hydrovatus clypealis Sharp
2.2-2.5 mm (page 100)

Plate 146. Family Dytiscidae
Hydrovatus cuspidatus (Kunze)
2.4-2.7 mm (page 100)

Plate 147. Family Dytiscidae
Hygrotus decoratus (Gyllenhal)
2.2-2.6 mm (page 103)

Plate 148. Family Dytiscidae
Hygrotus inaequalis (Fabricius)
2.7-3.5 mm (page 103)

Plate 149. Family Dytiscidae
Hygrotus quinquelineatus (Zetterstedt)
3.1-3.6 mm (page 104)

Plate 150. Family Dytiscidae
Hygrotus versicolor (Schaller)
3.1-3.6 mm (page 104)

Plate 151. Family Dytiscidae
Hygrotus confluens (Fabricius)
3.2-3.6 mm (page 104)

Plate 152. Family Dytiscidae
Hygrotus impressopunctatus (Schaller)
4.1-4.5 mm (page 105)

Plate 153. Family Dytiscidae
Hygrotus impressopunctatus lineellus
Gyllenhal
4.1-4.5 mm (page 105)

Plate 154. Family Dytiscidae
Hygrotus nigrolineatus (von Steven)
3.6-3.9 mm (page 105)

Plate 155. Family Dytiscidae
Hygrotus novemlineatus (Stephens)
3.5-4.0 mm (page 105)

Plate 156. Family Dytiscidae
Hygrotus parallellogrammus (Ahrens)
4.5-5.5 mm (page 105)

Plate 157. Family Dytiscidae
Hyphydrus aubei Ganglbauer
4.2-4.9 mm (page 106)

Plate 158. Family Dytiscidae
Hyphydrus ovatus (Linnaeus)
3.9-5.3 mm (page 106)

Plate 159. Family Dytiscidae
Laccornis oblongus (Stephens)
4.5-5.0 mm (page 106)

Plate 160. Family Dytiscidae
Laccophilus hyalinus (De Geer)
4.6-5.1 mm (page 108)

Plate 161. Family Dytiscidae
Laccophilus minutus (Linnaeus)
4.3-4.8 mm (page 108)

Plate 162. Family Dytiscidae
Laccophilus poecilus Klug
3.4-4.0 mm (page 108)